Ageing of Infrastructure

Ageing of Infrastructure

A Life-Cycle Approach

Frank Collins

Frédéric Blin

CRC Press
Taylor & Francis Group
Boca Raton London New York

CRC Press is an imprint of the
Taylor & Francis Group, an **informa** business

CRC Press
Taylor & Francis Group
6000 Broken Sound Parkway NW, Suite 300
Boca Raton, FL 33487-2742

First issued in paperback 2020

ISBN-13: 978-1-4665-8085-5 (hbk)
ISBN-13: 978-0-367-65703-1 (pbk)

Library of Congress Cataloging-in-Publication Data

Names: Collins, Frank (Engineer), author. | Blin, Frédéric (Engineer), author.
Title: Ageing of infrastructure : a life-cycle approach / Frank Collins
and Frédéric Blin.
Description: Boca Raton : Taylor & Francis, a CRC title, part of the Taylor &
Francis imprint, a member of the Taylor & Francis Group, the academic
division of T&F Informa, plc, [2018] | Includes bibliographical references
and index. |
Identifiers: LCCN 2018014612 (print) | LCCN 2018028976 (ebook) | ISBN
9780429849534 (Adobe PDF) | ISBN 9780429849527 (ePub) | ISBN 9780429849510
(Mobipocket) | ISBN 9781466580855 (hardback : acid-free paper) | ISBN
9780429455704 (ebook)
Subjects: LCSH: Weathering of buildings. | Building materials--Service life.
| Product life cycle.
Classification: LCC TH9039 (ebook) | LCC TH9039 .C65 2018 (print) | DDC
624.1028/8--dc23
LC record available at https://lccn.loc.gov/2018014612

Visit the Taylor & Francis Web site at
http://www.taylorandfrancis.com

and the CRC Press Web site at
http://www.crcpress.com

Contents

Preface

The book addresses the problem of ageing infrastructure and how ageing can reduce the Service Life to lower than expected. The rate of ageing is affected by the type of construction material, environmental exposure, function of the infrastructure and loading; each of these factors will be considered in the assessment of ageing. How do international design codes address ageing? Predictive models of ageing behaviour are available, and the different types (empirical, deterministic and probabilistic) will be discussed in a whole-of-life context. Life-cycle plans, initiated at the design stage, can ensure that the Design Life is met, while optimising the management of the asset – reducing life-cycle costs and reducing the environmental footprint due to less maintenance/remediation interventions and less unplanned stoppages and delays. Health monitoring of infrastructure can be conducted via implanted probes or by non-destructive testing or *by non-contact wireless methods* that can routinely measure the durability, loadings and exposure environments at key locations around the facility. Routine monitoring can trigger preventative maintenance that will extend the life of the infrastructure and minimise unplanned and reactive remediation, while also providing ongoing data that can be utilised toward more durable future construction. Future infrastructure will need to be safe and durable, financially and environmentally sustainable over the life cycle, thereby raising the socioeconomic well-being – the book concludes by discussing the key impacting factors that will need to be addressed. The authors have a strong academic and industry background. The book will be a resource for academics and practitioners wishing to address ageing of built infrastructure.

Professor Frank Collins
Deakin University

Dr Frédéric Blin
AECOM Ltd

About the authors

Frank Collins is the director of the Australian Centre for Infrastructure Durability (ACID). His role involves fostering collaboration between universities to streamline access by government and industry to the most relevant researchers and facilities for the durability of built infrastructure. Prior to his academic career, Frank has had nineteen years as a chartered professional engineer, which gives him a unique perspective on ageing of infrastructure. He worked for thirteen years with Taywood Engineering Ltd in their London, Hong Kong, Sydney and Melbourne offices, involved with diagnosis and rehabilitation of infrastructure and applications of construction materials. In 1988, Frank was lead materials engineer in the Sydney Opera House rehabilitation program, which entailed diagnosis of the roof shell elements and substructure, and development of remedial and preventative maintenance works. In 1995, he established the Bridge Testing and Rehabilitation Unit within the Ministry of Transport, Vietnam, allowing the Ministry to become self-sufficient in the management of bridge assets. While technical director at AECOM (Maunsell), he established and managed the company's Advanced Materials Group, including technical and commercial leadership, and provided high-level advice on many international infrastructure projects.

As an academic since 2006, he has taken a keen research interest in the areas of durability and ageing of built infrastructure, utilisation of wastes as alternative construction materials and improved construction materials for durable and stronger infrastructure.

Frédéric Blin has over eighteen years' experience as an engineer and is the leader of AECOM's Strategic Asset Management and Advanced Materials team in Victoria, Tasmania and South Australia. He has worked across industries (properties/facilities, transport and water) and brings a solid understanding of the importance of balancing asset life-cycle risks and costs to provide desired service levels to customers. He holds a materials engineering degree and a PhD in corrosion engineering.

At AECOM, Frédéric is a technical director, who has worked on and led numerous projects in the field of asset management and durability/materials engineering. This has included the provision of technical support at the design, build, finance (including due diligences), operation and maintain phases of the life cycle of (particularly infrastructure based) assets (e.g. parks, tunnels, ports, plants, buildings, roads and bridges).

Frédéric's experience includes condition and performance monitoring, evaluation, risk management, life-cycle decision frameworks and tools, maintenance and renewals forecasting, as well as asset management business improvement.

Chapter 1

Introduction

WHAT IS BUILT INFRASTRUCTURE?

View the world around you. Built infrastructure is vast. It sustains our quality of life: homes provide shelter; buildings accommodate a multitude of activities; transport infrastructure connects us via our roads, bridges, airports, railways and ports; built services provide for our well-being (energy generation and distribution, communication nodes and transmission, water and wastewater storage, treatment and reticulation). The management of our ageing infrastructure assets throughout the life-cycle (from planning, construction and utilisation to demolition) has had, and will continue to have, important social, economic and environmental impacts. It is within this framework that greatly influences our relationships within social, commercial and environmental contexts. As a result, it is essential to understand how these assets were designed and built, including the materials utilised during construction (Figure 1.1).

CONSTRUCTION MATERIALS AND METHODS OF CONSTRUCTION

The range of construction materials used to build our infrastructure is large, as choices are often influenced by availability and economy. The most utilised material is concrete, with 25 gigatonnes per year consumed globally (Gursel et al., 2014) on the construction of built infrastructure. Infrastructure is also comprised of a range of steel and alloyed metals utilised as structural frames, pipelines, cladding, fixtures, organic solids (e.g. polymers – including plastics, composites and sealants), masonry, timber and glass. Each has unique properties and each has different levels of resilience depending on exposure conditions and service loadings.

Figure 1.1 Built infrastructure is vast and varied, comprising transportation, maritime, buildings, industrial, energy, water and wastewater needs – all exposed to different loading and exposure conditions.

AGEING

The broad range of materials utilised to construct and maintain infrastructure is exposed to specific environments during the asset's life. Depending on location and function, the exposure environment imposes chemical and physical actions – both in terms of the overall geography (climate and service loadings) and microclimates (localised exposures and functionality). All materials wish to return to a lower energy state, and manufactured construction materials have locked-in energy that through the laws of nature will return eventually to the original state: metals corrode, timber rots, concrete deteriorates and polymers become brittle. Ageing can be subtle and not easily recognised, and the impacts of deterioration may involve the critical-built components that could lead to premature failure. The rate of ageing is affected by the type of construction material, environmental exposure, deterioration mechanism (also referred to as failure modes) and loading. Population increases have led to greater demands and higher loadings and more frequent use, causing greater impact on ageing. Modern infrastructure is built to last a desired Design Life before they are upgraded, renewed or replaced. If premature ageing occurs, then this could have significant consequences for their users and societies as a whole.

HISTORICAL CONTEXT

Historically, built infrastructure has played a key role since ancient times. Most noticeable are the pyramids of Giza (Nell and Ruggles, 2014), arched bridges, elevated roads and aqueducts across valleys. Roman aqueducts carried water over long distances in order to provide a crowded urban population with relatively safe, potable water and sewers (Wolfram and Lorenz, 2016); transport networks (Carreras and de Soto, 2013) and ancient Mediterranean harbours (Marriner and Morhange, 2007). All infrastructure ages, although in the ancient context it must be born in mind that historic infrastructure is currently maintained with considerable conservation effort and cost (Figure 1.2).

The overarching aim of modern infrastructure is to strike a balance between the lifecycle risks and costs of these assets and the services they provide over time. The large population growth over the last century coupled with changing climatic conditions has put a significant stress on our infrastructure. This has resulted in the need to adapt these assets to current and future demands as well as develop adaptable and resilient new infrastructure solutions.

Figure 1.2 Aqueduct arcade in the Vallon des Arcs on the Barbegal System in Southern France. Construction of the arches is primarily of rubble, mass concrete, with a brick-and-mortar skin. (From Wolfram, P.J., and Lorenz, W.F., *Int. J. Hist. Eng. Technol.*, 86, 56–69, 2016. Taylor & Francis Group.)

BUILT FOR A FINITE LIFE

Built infrastructure is typically designed for a defined life. In Australia, steel and reinforced concrete structures are designed for fifty-year life (Standards Australia, 2009) or, in the case of bridges, one hundred years (Standards Australia, 2004). Design standards vary globally depending on local needs and exposure conditions and functionality. However, the reality is that significant portions of railway infrastructure in many cities is *over* one hundred years old – is it realistic to replace this infrastructure because one hundred years has expired? Demolition and reconstruction is energy intensive, generates pollution and noise, causes disconnections/delays to public services, and associated costs, whereas simpler methods of restoration may be available.

IMPACT OF AGEING

Impact of ageing on built infrastructure and on the social and economic fabric is significant. There have been examples where the rapid and unpredicted ageing of infrastructure elements led to structural failure resulting in the loss of lives, injuries and environmental damage as well as major reconstruction costs. A key case relates to the Minneapolis bridge collapse in 2007 where 13 people were killed, 145 injured, causing major disruption to a key transportation system, and the cost of a replacement bridge costing over $300 million (Nunnally, 2011). The cost of maintaining and rehabilitating built infrastructure is significant, expending approximately 3.8 percent of global GDP (Gursel et al., 2014). In 2017, the American Society of Civil Engineers Infrastructure Report Card provided a cumulative rating of D+, 'poor or at risk', and an estimated $206 billion each year is needed to rehabilitate US infrastructure from 'poor' to 'adequate' condition (ASCE, 2017). The direct impacts of ageing infrastructure are often very visible through rectification/replacement and the resources needed to undertake remedial work (e.g. the premature reconstruction of a deteriorated bridge that has not achieved the full Design Life). Another example relates to the environmental cost arising from oil seepage in oilfields due to corroded pipelines. Nevertheless, the effects of ageing of buried, underwater and difficult-to-access infrastructure are less visible, and the problems can be insidious.

As shown in Figure 1.3, the outcomes from aged infrastructure are far-reaching and can affect a broad range of industry and public sectors. Deteriorating infrastructure reduces functionality, safety and efficiency of existing buildings (domestic, commercial, industrial and heritage), transportation infrastructure, mining, defence, energy, communications and water/wastewater infrastructure. However, the flow-on economic/environmental/social effects of ageing are less understood. Research on the degradation of

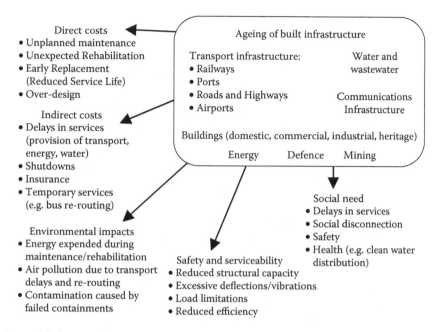

Direct costs
- Unplanned maintenance
- Unexpected Rehabilitation
- Early Replacement
  (Reduced Service Life)
- Over-design

Indirect costs
- Delays in services
  (provision of transport,
  energy, water)
- Shutdowns
- Insurance
- Temporary services
  (e.g. bus re-routing)

Environmental impacts
- Energy expended during
  maintenance/rehabilitation
- Air pollution due to transport
  delays and re-routing
- Contamination caused by
  failed containments

Ageing of built infrastructure

Transport infrastructure: Water and
- Railways   wastewater
- Ports
- Roads and Highways   Communications
- Airports   Infrastructure

Buildings (domestic, commercial, industrial, heritage)

Energy   Defence   Mining

Social need
- Delays in services
- Social disconnection
- Safety
- Health (e.g. clean water
  distribution)

Safety and serviceability
- Reduced structural capacity
- Excessive deflections/vibrations
- Load limitations
- Reduced efficiency

Figure 1.3 **Ageing of built infrastructure showing the broad types of infrastructure that are utilised by the public, industry and government as well as the direct and flow-on impacts. (Courtesy of F. Collins.)**

engineering materials has received most attention over many years, driven by the science of materials behaviour and modelling of deterioration rather than consideration of the impacts on our quality of life. It is difficult to quantify the societal effects of delays, shutdowns, congestion, energy risk and social dislocation caused by failed infrastructure. Similarly, the flow-on economic effects to, say, an aged port structure, could lead to reduced capacity, productivity, integrity, serviceability and associated reduced road/rail clearances from/to the port. The economic costs due to delayed imports and exports of goods, delivery of services and dependant businesses is substantial. The follow-on effects to the environment are also difficult to quantify but can be affected by road congestion leading to poorer air quality, contamination of ground, waterways and air by leaking containments and pipelines, and the carbon dioxide emissions associated with the energy expended during rehabilitation and premature reconstruction of deteriorated infrastructure. Flow-on social impacts relate to the aforementioned factors as well as dislocation due to transportation delays, reduced safety, impaired health from leakages of contaminants and impacts of delays and shutdowns of energy, communications and water/wastewater services to homes and public facilities.

KEY ISSUES TO BE ADDRESSED IN THIS BOOK

New methodologies for the design as well as forensic examination and rectification for infrastructure are moving rapidly with the advent of new technologies. The later chapters of this book will deal with these items, while the important basics will be covered in Chapters 1 through 4. Chapter 2 will deal with the common argument; the comparison between Design Life with Service Life and where ageing impacts on both. The mechanisms of ageing are covered comprehensively in Chapter 3. The exposure environment, covered in Chapter 4, plays a significant role in the deterioration of construction materials and therefore impacts on ageing. In Chapter 5, predictive modelling of ageing is critical during the design stage and also during the Service Life to enable streamlining of maintenance. Whole-of-life Engineering for Ageing Infrastructure is addressed in Chapter 6 and follows the life of the infrastructure from concept design, detailed design, construction phase, through to post-construction Service Life. Health monitoring of built infrastructure has changed enormously from virtual and hand-held nondestructive testing to wireless, noncontact methods of harnessing large volumes of data and decision-making related to many properties and attributes of the infrastructure, including ageing. Much of this is presented in Chapter 7. Chapter 8 provides a review of future materials, methods, the impacts and future means of dealing with ageing infrastructure.

REFERENCES

ASCE (2017). American society of civil engineers report card. https://www.infra-structurereportcard.org.

Carreras, C. and de Soto, P. (2013). The Roman transport network: A precedent for the integration of the European mobility, *Historical Methods*, 46(3): 117–133.

Gursel, P. A., Masane, T. E., Horvath, A., and Stadel, A. (2014). Life-cycle inventory analysis of concrete production: A critical review, *Cement and Concrete Composites*, 51: 38–48.

Marriner, N. and Morhange, C. (2007). Geoscience of ancient Mediterranean harbours, *Earth Science Reviews*, 80: 137–194.

Nell, E. and Ruggles, C. (2014). The orientations of the Giza pyramids and associated structures, *Journal for the History of Astronomy*, 45(3): 304–360.

Nunnally. (2011). *The City, the River, the Bridge: Before and after the Minneapolis Bridge Collapse*. Minneapolis, MN: University of Minnesota Press, 216p.

Standards Australia. (2004). *AS5100 Bridge Design, Part 1: Scope and General Principles*. Sydney, Australia: SAI Global, 49p.

Standards Australia. (2009). *AS3600 Concrete Structures*. Sydney. Australia: SAI Global, 213p.

Wolfram, P. J. and Lorenz, W. F. (2016). Longstanding design: Roman engineering of aqueduct arcades, *International Journal for the History of Engineering and Technology*, 86(1): 56–69.

Chapter 2

Contrasting Design Life with Service Life – effects of ageing

INTRODUCTION

When considering the life of built infrastructure, the terms Design Life and Service Life are commonly used and incorrectly exchanged for the same meaning. Both terms refer to the length of time that the built item will perform for its intended purpose; the key difference between the terms is *intended* versus *actual* performance. Design Life defines the intended life of the infrastructure and without major repair being necessary. Design life occurs prior to construction, whereas Service Life and environmental exposures that contribute to ageing. Premature replacement of built infrastructure before expiry of Design Life due to ageing highlights the importance of these two terms (Figure 2.1). This chapter discusses these terms within the context of ageing and considers life cycle implications.

DESIGN LIFE

Ageing of infrastructure has drawn attention to how Design Life is considered by designers, builders and asset owners/managers *prior* to construction. It takes on a range of definitions and specified durations, depending on the type, function, operating environment, prestige, vulnerability and criticality of the built infrastructure. For example, the Design Life of an offshore oil structure is a moving target, depending on changes in loading during the operating life, oil and gas price fluctuations and projected reserves of oil versus actual.

The baseline definition adopted by design codes is the period for which the built infrastructure is required to perform its intended purpose (JSCE, 2010), although most design codes further stipulate *shall be constructed that all the design, as contained in the drawings and specifications, are achieved*, as advocated in the Australian Standard AS3600 Concrete Structures (Standards Australia, 2009). This includes serviceability, fatigue and fracture, strength and extreme events (e.g. earthquakes). This places greater onus on the designers and builders to incorporate suitable

Figure 2.1 The cracking of a column due to chloride-induced corrosion caused by con-
taminated aggregates and mixing water. The age of the structure at first cracking was only three years old from the completion of construction. (From
Rostam, S., Chapter 1, in C. W. Yu and J. Bull (Eds.), *Durability of Materials and
Structures in Building and Civil Engineering,* CRC Press, 462pp, 2006.)

durability within the built infrastructure. The definition has broadened in
recent years, for example ISO 16204 (ISO, 2012), to include the obligation
of the asset owner of the built infrastructure to undertake regular maintenance to enable the designed life to be met during service, but without
major rehabilitation.

Infrastructure owners may further stipulate that Design Life also fulfils
particular needs. Examples include: limits on embodied carbon (Basbagill
et al., 2013); environmental protection as stipulated by the Canadian
Nuclear Safety Commission (CNSC, 2008) and financial viability over the
duration of the Design Life (Zhang, 2005).

The duration of Design Life of a standard bridge in Australia is one hundred years (AS5100, 2017). However, in the specific case of the Gateway
Bridge in Brisbane (Australia), together with 20 km of freeway upgrade,
the Queensland Department of Main Roads specified *three hundred years*
Design Life due to the criticality of the infrastructure (Hart and Connal,
2012). The completed bridge is shown in Figure 2.2. During design, a
comprehensive Durability Plan was prepared, setting out for each designed
component the identification of deterioration mechanisms, including
materials selection; the development of mitigation measures to ensure that
the design intent is met; the identification of durability critical construction activities; verification of constructed components to confirm compliance with durability requirements; protective measures implemented in

Figure 2.2 Gateway bridge duplication designed for three hundred years.

the project; inspection and maintenance provisions and record of ongoing activities during the Service Life pertaining to durability.

Durability Plans are common and becoming mandated within the design of major infrastructure. The exposure conditions are varied and prone to deteriorate the completed structure, for example, the Gateway Bridge in Brisbane, Australia:

1. Components located within the soil are vulnerable to corrosion of exposed steel/reinforced concrete due to, for example, exposure to saline groundwater, stray current corrosion due to proximity to high-voltage sources, deterioration of concrete due to exposure to sulphates or the exposure to acid-sulphate soils containing pyrite, which transforms to corrosive sulphuric acid if exposed to air (oxygen) during excavation.
2. Above-ground risks to exposed structural elements including corrosion due to exposure to airborne chloride and carbonation (acidic gases combined with water vapour) as well as exposure to ultraviolet light, dewfall and rainfall.
3. Water-borne components are subject to erosion and scour, abrasion by debris and corrosive chloride.

Durability features of the Gateway Bridge construction included: education programmes for all workers on the importance of '3C compaction-cover-curing' of concrete; selective use of stainless steel reinforcement in the corrosive splash zone and provisions for future cathodic protection; utilisation of high quality concrete; application of protective coatings to concrete surfaces as well as to hot dip galvanised structural steelwork. The consideration of some of the variability associated with the issues earlier were accommodated by a stochastic approach considering the variability expected on environmental conditions and the variability of the material properties and adopting a reliability approach to durability design. An asset management plan was developed for the ongoing operation of the freeway, including provisions for ongoing access for inspection, testing and maintenance.

The type of infrastructure influences the Design Life: a maritime structure has a Design Life requirement of fifty years (Standards Australia, 2005) compared with one hundred years for a bridge (AS5100). Depending on specific client needs, these durations may be expanded. Bridges are critical components of road infrastructure, and removing a bridge from service would attract direct costs for replacement and additional flow-on costs (delays in services to the public and industry) and social dislocation. While maritime structures are also important to the transport network, the fifty-year Design Life is more appropriate to the anticipated changes in port function and layout and the rigours of the corrosive marine environment. Past methods of design for durability have been based on 'deemed to satisfy' methods regarding durability of materials subject to environmental exposures. Nevertheless, recent approaches are now based on intended life that can be achieved with an unacceptable level of reliability (to be discussed in the next section).

SERVICE LIFE

Service Life begins at the completion of construction and ceases at the time of decommissioning/demolition or failure of the infrastructure. The problem of ageing leading to premature replacement of built infrastructure before expiry of Design Life has a legacy of economic, environmental and social impacts. If ageing occurs at an advanced rate, Service Life could be considerably less than Design Life. Conversely, a very durable infrastructure could exceed the Design Life (as defined in ISO 13823). Planned obsolescence may occur when the functionality of the infrastructure changes during the Service Life, for example with the change of operation of a port from a bulk liquids berth to containerisation would lead to out-dated facilities.

Disparities exist between the predicted versus actual service performance of built infrastructure. Weaknesses in traditional codes have included: lack of clarity as to what constitutes end of Service Life; level of necessary reliability and lack of long-term data based on field experience. Limit-state-based Service Life design recognises the need to be specific about what condition represents the 'end of Service Life'. Alternative design approaches consider the life cycle aspects of the built infrastructure (FIP, 2006 and ISO, 2012).

FIP Bulletin 34 (FIP, 2006) was compiled by Task Group 5.6; experts from Europe, North and South America and Japan based the approach to Service Life on design based on a limit state (LS) and reliability-based concept. Service Life Design (ISO, 2012) considers plain concrete, reinforced concrete and pre-stressed concrete structures with a special focus on design provisions for managing the adverse effects of degradation (FIP, 2006; ISO, 2012) and outlining the basis for Service Life design, providing four different options:

1. A full probabilistic approach based on characteristic values of actions and materials properties based on data derived for the particular project or from general field experience or from relevant literature
2. A semiprobabilistic approach (partial factor design) where the durability uncertainties are engaged by applying a Safety Factor to representative properties of actions and materials
3. Deemed to satisfy rules
4. Avoidance of deterioration

The code also provides for serviceability limit states that allow for local damages to constructed components or changes in the function of a particular component (e.g. excessive vibrations or visual deterioration of nonstructural components).

The distinctive shell structures comprising the Sydney Opera House (Figure 2.3) were built to create an architectural landmark and state-of-the-art venue for the performing arts in the city of Sydney.

Figure 2.3 Sydney Opera House. (Courtesy of Thomas Adams, 2016.)

Figure 2.4 Frank Collins (Author) inspecting deteriorated elements of the Sydney Opera House roof shell. (Courtesy of F. Collins.)

In 2007, UNESCO approved the Sydney Opera House to be inscribed on to the World Heritage List. Prominent at Bennelong Point within Sydney Harbour, it provides a magnificent vista; however, the location is vulnerable to ongoing exposure to the corrosive marine environment as well as the relentless cycles of weather. The prestigious structure was designed for two hundred and fifty years, but only fifteen years after the official opening, major rehabilitation works were needed. The broadwalk support structure, located in marine tidal and splash conditions, exhibited widespread steel reinforcement corrosion within the reinforced concrete A-frame and strut elements (Collins, 1990a; Pratley 1990) leading to major rehabilitation in 1996 by cathodic protection (Tettamanti et al. 1997). The roof shells also needed a comprehensive visual survey together with non-destructive testing using specialised abseiling methods of access (Figure 2.4).

The types of defects that were identified included detachment of edge tiles, deterioration of tile-to-tile sealants, elastomeric sealant integrity between adjacent concrete panels and isolated areas showing reinforcement corrosion (Collins, 1990b).

SECOND-GENERATION INFRASTRUCTURE

Construction and operation of infrastructure serves social, economic and environmental needs; therefore, the demolition and replacement have major adverse impacts. Building elements are often designed with highly

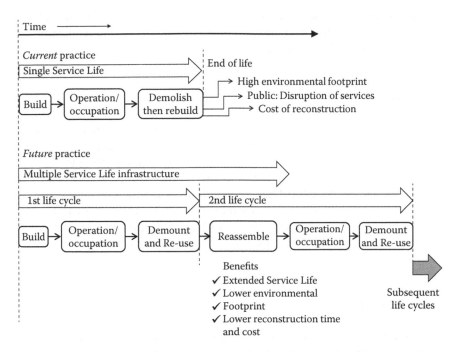

Figure 2.5 Consideration of multiple service lives of infrastructure. Ageing of different components and materials will need clever engineering solutions. (Courtesy of F. Collins.)

interdependent components, and therefore most construction is monolithic, necessitating demolition and replacement at the end of the Service Life. Hence, it is virtually impossible to take one part of the assembly apart without affecting the structural system. If construction could be considered with a view toward reconstruction at the end of the Service Life, then the social, economic and environmental benefits could be significant (Figure 2.5).

Construction that allows for future demounting and reassembly at the end of the Service Life (or when changes of function govern the need to modify the structure) has advantages over conventional 'demolish then rebuild' (Reinhardt, 2012). There is a pressing need to design assemblies with connections that allow each part to be replaced discretely, recognizing the very different Service Life time spans of different built components. Further, the ageing of the various components that comprise a built facility must be considered for longevity, maintenance and/or replacement for a second or multiple service lives.

This type of approach is not confined to buildings; it is also applicable to bridges, port structures, energy infrastructure, telecommunication structures and other built facilities.

CONCLUSION

Design Life and the consideration of ageing during the design and construction stages are critical for construction of durable infrastructure. Premature deterioration can lead to a Service Life that impedes the designed life and excessive maintenance, rehabilitation or, in the worst case, replacement before the Design Life has been served. This leads to economic, environmental and social consequences. Durability Plans that consider the life cycle durability, or the construction of multiple life components/structures, are encouraging signs that mitigation of ageing is being considered for future infrastructure.

REFERENCES

Basbagill, J., Flager, F., Lepech, M., and Fischer, M. (2012). Application of life-cycle assessment to early stage building design for reduced embodied environmental impacts, *Building and Environment*, 60: 81–92.

CNSC. (2008). RD–337 design of new nuclear power plants, Canadian Nuclear Safety Commission (CNSC), 77p. http://nuclearsafety.gc.ca/eng/acts-and-regulations/regulatory-documents/published/html/rd337/.

Collins, F. (1990a). The Broadwalk: Investigations, Sydney Opera House upgrade–The structure and its maintenance, *Proceedings of Joint Seminar Organised by the Institution of Engineers*, Australia and the Concrete Institute of Australia, pp. 50–55.

Collins, F. (1990b). The Roof: Investigations, Sydney Opera House upgrade–The structure and its maintenance, *Proceedings of Joint Seminar Organised by the Institution of Engineers*, Australia and the Concrete Institute of Australia, pp. 1–8.

FIP. (2006). Model code for service life design, *International Federation for Structural Concrete*, Task Group 5.6, Bulletin 34, 116p.

Hart, J. and Connal, J. (2012). 300-year sustainability: The second gateway bridge Brisbane, *Proceedings NZ Bridges*, Wellington, New Zealand, ARRB Group Limited, 6p.

ISO. (2012). *ISO 16204 Durability–Service Life Design of Concrete Structures*. Geneva, Switzerland: International Standards Organisation, 31p.

JSCE. (2010). Standard specifications for concrete structures–Design. *JSCE Guidelines for Concrete 15*, 503p.

Pratley, J. (1990). Sydney Opera House upgrade–Investigation and repair of the substructure, *CIA News*, 16(4): 12–15.

Reinhardt, H. W. (2012). Demountable concrete structures–An energy and material saving building concept, *International Journal of Sustainable Materials and Structural Systems*, 1(1): 18–28.

Rostam, S. (2006). Chapter 1, In C. W. Yu and J. Bull (Eds.), *Durability of Materials and Structures in Building and Civil Engineering*, Dunbeath, Scotland: CRC Press, 462p.

Standards Australia. (2005). *AS4997 Guidelines for the Design of Maritime Structures*, Sydney, Australia: SAI Global, 57p.

Standards Australia. (2009). *AS3600 Concrete Structures. Part 1 Scope and General Principles*, Sydney, Australia: SAI Global, 53p.

Standards Australia. (2017). *AS5100 Bridge Design, Part 1 Scope and General Principles*. Sydney, Australia: SAI Global, 53p.

Tettamanti, M., Rossini, A., and Cheaitani, A. (1997). Cathodic prevention and Cathodic protection of new and existing concrete elements at the Sydney Opera House, *Proceedings of Corrosion 97*, 255/1–9. Sydney, Australia: Australasian Corrosion Association

Zhang, X. (2005). Financial viability analysis and capital structure optimization in privatized public infrastructure projects, *Journal of Construction Engineering and Management*, 131(6): 656–668.

Chapter 3

Mechanisms of ageing

INTRODUCTION

Manufacturing of most construction materials involves considerable energy input to transform the raw materials into a processed item. In the case of steel, conversion of iron ores to steel is an energy-intensive multiphased manufacturing process, resulting in a construction product that is strong and resilient but nevertheless thermodynamically unstable. Slowly with the elapse of time, chemical reactions between the steel and the surrounding environment cause deterioration (corrosion), releasing energy and reverting the manufactured steel eventually to its natural state: iron oxide, or 'rust' that is characterised by section loss and loss of performance. Over the life of a constructed component and following exposure to the external environment, metals corrode, timber rots, concrete degrades and polymers become brittle, for instance. Ageing is reflected by responses by individual materials within built components, while contributing to the ageing of the overall infrastructure over a period of time. This chapter examines why engineering materials age and what are the key ageing mechanisms in the context of built infrastructure.

MACRO- TO NANO-SCALE

When characterising the ageing behaviour of built infrastructure, a multi-scale understanding of a built component is needed. For example, concrete that comprises a bridge, as illustrated in Figure 3.1, consists of many components, and a key material compromising the bridge is reinforced concrete. The concrete, in turn, consists of fine and coarse rock aggregates that are embedded within a cementitious binder. Figure 3.1 leads to the nano-scale interactions of the physical and chemical makeup of the constituent materials: from the atomic-size to larger-size ranges (and vice versa). The quality of the raw materials and fabrication are key factors at the macro-scale, including the design, specification, workmanship and

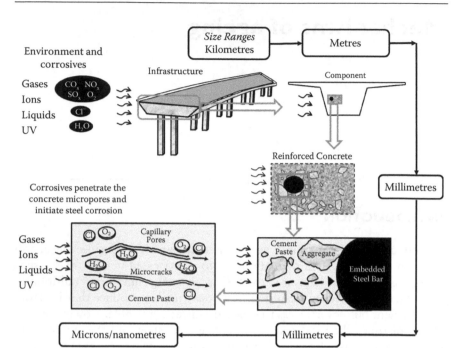

Figure 3.1 Multi-scale considerations for assessment of durability. (Courtesy of F. Collins.)

quality of the manufactured component that also greatly influence the rigour of the built structure to resist ageing.

Case example: size range scaling effects

Figure 3.1 illustrates a large bridge comprised of many components and subcomponents: this example will focus on the beams that form the super-structure. In a benign environment, and assuming no prior mechanical damage, reinforced concrete (RC) remains durable, particularly the embedded steel, which remains passive due to the high alkalinity of concrete.

In Figure 3.1, the bridge resides in a coastal and industrial/urban location; hence, the exposure of the beams is primarily to airborne chloride and acidic gases (including airborne carbon dioxide, nitrous and sulphurous oxides), wind-driven rainfall, water vapour and ultraviolet light. The beams are vulnerable to exposure to these agents, and the porous nature of concrete facilitates entry of these agents, eventually to the depth of the embedded steel that is vulnerable to corrosion. The contributions of scale are critical when considering ageing. The overall structure is of the order 10^2 to 10^3 metre size range and a single beam component of the bridge, shown in Figure 3.1, is within the order 10^1 to 10^2 metre size range. When a slice of the beam component is examined within the metre to millimetre range, Figure 3.1 shows a steel bar embedded in the concrete, and although the surrounding concrete provides protection (termed 'cover'), the surrounding concrete is nevertheless a permeable medium to corrosives. At the nano- to micro-scales, the corrosives slowly migrate into the pervious concrete. Concrete is porous for two reasons:

1. Cement paste has a complex microstructure consisting of hydration products (gel), unreacted cement and pores. During hydration, gel gradually replaces water-filled space, forming impermeable gel pores, and the unfilled water voids become capillary pores. Capillary porosity depends on the ratio of water to cement (w/c) while gel pores are smaller and disconnected and therefore not considered as viable for the ingress of corrosives. Larger capillary pores and microcracks facilitate transport of corrosives.
2. Microcracks form at the weak aggregate-cement paste transition zone, a zone prone to poorer packing, separation of water from cement leading to higher w/c (higher porosity) and precipitation of weak calcium hydroxide crystals. Even at low stress levels, microcracks emanate at this zone due to restraint of the weaker cement paste by the stiffer coarse aggregates during autogeneous shrinkage. As the loading on the bridge increases, loading induces the further formation of more microcracks, while the tensile strains produced by concrete shrinkage produce cracks, some connecting with capillary pores and eventually connecting to create a crack.

PATHWAYS TO AGEING

The constituents of porous materials are physically and chemically complex. Nevertheless, the aforementioned conduits in concrete provide the primary pathways for corrosives to travel. The mechanisms for corrosives transport into the concrete can be different, for example, in the case of Figure 3.1:

1. Leaching of hydroxyl ions from the concrete when in contact with lower pH water or low-calcium water. This can be influenced by whether the contact is constant or intermittent, and also the structural concrete member type and geometry and exposure type. Cumulative leaching, in turn, affects the pH of the pore water inside the concrete as well the outflux water (Law and Evans, 2013). In flowing water, the speed and type of flow, whether laminar or turbulent, can also affect the rate of leaching, which will, in turn, affect the pH of the pore water inside the concrete. This is problematic

because the high pH of the concrete gradually reduces to a level 9.0–9.5 that is sufficient to depassivate the embedded steel reinforcing and initiate corrosion.

2. Rainwater becomes absorbed into the outer exposed concrete via convection, up to tens of millimetres in very porous concrete, while assisting the carriage of airborne corrosive chloride into the concrete. A key mechanism is capillary suction (sorption) in an empty or partly saturated capillary pore network. A common theoretical model, which relates time and the height or depth of penetration of a liquid into an empty capillary, is the Lucas Washburn equation (Hall and Tse, 1986; Collins and Sanjayan, 2008; Lucija et al., 2010):

$$S = \frac{i}{t^x} \tag{3.1}$$

where:
S is the sorptivity coefficient (mm/min$^{0.5}$)
i is the hydraulic diffusivity Q/A where Q is the amount of water adsorbed (mm^3)
A is the area exposed to water
t is time
x is a constant that applies to different porous materials (e.g. $x = 0.5$ is generally applied to concrete)

3. Diffusion of chloride (and water) following a concentration gradient (e.g. high chloride at the exposed surface to low concentrations inside the concrete).

4. When chlorides, or chloride/hydroxide, finally reach above a threshold concentration at the depth of the steel bar, the passive iron oxides transform into expansive corrosion products: 'rust'. Loss of section and tensile pressure caused by the expanding rust causes the outer concrete to crack and detach from the structure.

5. Airborne acidic gases react with water vapour inside the concrete, forming acids, which progressively neutralise the protective alkalinity (approximately 12.5) of the concrete. When the pH of the concrete is reduced to approximately 9.0, the passive iron oxide on the steel transforms into a corrosive state, leading to section loss and expansive oxides: as described in (2).

6. Chemical reactions between the cement paste and reactive chemicals (e.g. spillages on the deck by acids, sulphates, decalcification and ammonium nitrate) (Figure 3.2).

To understand the deterioration of infrastructure, we need to consider the many mechanisms of ageing across a range of engineering materials and behaviour across a multi-scale ranging from metres to nanometres.

Figure 3.2 Concrete plinth deteriorated by sulphate (chemical) attack. (Courtesy of F. Collins.)

MECHANISMS OF AGEING

Ageing is generally a slow process and, therefore, unlike an earthquake, it seldom features prominently in the news media unless, following years of ageing, there is a major calamity, such as the sudden collapse of the I-35W Mississippi River Bridge in Minneapolis that killed thirteen people on 1 August 2007 (National Transportation Safety Board, 2008.).

Figure 3.3 summarises the mechanisms of ageing that are encountered by a broad range of infrastructure types, construction materials and service conditions. As stated earlier, accelerated deterioration, including fire and explosion, although included in Figure 3.2, are not the focus of this book, which instead is the gradual ageing of infrastructure, which is more common, considerably slower and exhibits progressive loss of function until finally becomes obsolete or fails to meet social economic and/or financial expectations.

For most engineering materials, depending on the type, composition and exposure environment, it is common that several of the mechanisms shown in Figure 3.3 will act simultaneously or intermittently. For example, a timber wharf can be exposed to the simultaneous actions of abrasion by regular wave impacts, loading and unloading over the lifecycle (mechanical), regular

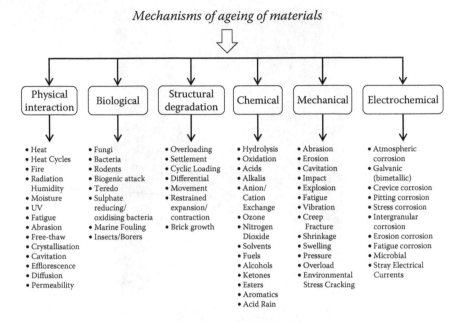

Figure 3.3 Mechanisms of ageing of a vast range of engineering materials. (Courtesy of F. Collins.)

cycles of UV sunlight, temperature, tides, various biological exposures (e.g. teredo worm/borer or fungi) and corrosion of fixtures, such as bolts and steel bracing (electrochemical). Furthermore, most built facilities are composed of a range of different engineering materials (e.g. concrete façade panel fixed to a steel-framed building), or composites (e.g. fibre-reinforced composites are a combination of fibrous materials embedded within a resin-based matrix). It is difficult to generalise for every type of material and service type; however, Figure 3.3 provides a straightforward reference toward classifying the broad range of mechanisms toward deterioration, including plastics, paints, timber, concrete, ceramics and metals. Each mechanism is discussed as follows:

1. Physical interaction
2. Biological
3. Structural degradation
4. Chemical
5. Mechanical
6. Electrochemical

Physical interaction

A broad range of physical actions degrade engineering materials. Examples include heat, impact, attrition, structural movement and expansion/contractions due to temperature fluctuations.

Abrasion

Abrasion is caused by solids and/or liquids mechanically rolling or sliding over a material thereby leading to wear. Examples of abrasion damage are visible on marine infrastructure (exposed to wave splash or wind-blown sand), wear on trafficked roads and industrial floors and erosion of spillways on hydraulic structures.

Freeze-thaw

Freeze-thaw leads to damage of porous materials (e.g. stone masonry and concrete) located in cold climates, and it is characterised by saturated water within the capillary pores being exposed above and below the freezing point. Water expands approximately 9 percent when frozen to a solid, and cycles of freezing/thawing leads to damage to the pores. In addition, the sliding action of frozen/unfrozen water further damages the capillary walls due to attrition and eventually causes cracks and detachment of the exposed surface. If the freezing liquid is saline, the exposed material becomes flakey at the surface as a result of salt scaling.

Crystallisation

Porous materials, including masonry, stone and concrete, are vulnerable to damage due to salt crystallisation in the exposed pores (Zhutovsky and Hooton, 2017). Cyclic weathering and repeated wet/dry exposures can generate salt crystals that eventually exceed the tensile capacity, leading to failure of the exposed material. The main types of crystals include sodium sulphate, sodium chloride and sodium carbonate. This is a cyclic process within the pores, salt crystallises out as water evaporates due to a decrease in the ambient RH, and the salt will redissolve as the RH is increasing again. Salt crystals growing within confined pore spaces generate disruptive stress, progressively causing damage that renders surface deterioration of the surface of porous building materials. The extent of damage depends on the supersaturation of the salt, proximity to the wetting/drying front, the size of the pores and a disjoining pressure between the growing crystal and pore wall.

Temperature

Polymers are vulnerable to heat exposure (Burks and Kumosa, 2012) and possess a glass transition temperature (T_g); i.e. if the temperature of the polymer exceeds T_g, the polymeric properties are transformed from rigid, elastic and glassy into a softer, weaker viscoelastic material (Summerscales, 2014). The chemical processes involved during degradation can influence the chemical composition of the polymer and also the physical parameters, including chain conformation, molecular weight, molecular weight distribution (MWD), crystallinity, chain flexibility, cross-linking and branching. A synergistic effect of high temperature and solar UV radiation can cause rapid degradation of polymers such as polyethylene.

Permeability

Plastics are not impermeable to gases and liquids: (1) Depending on molecule size, a passage that can be utilised is by passing through a small hole or leak; (2) Nano-cracking allowing the passage through spaces between molecules or atoms in the crystal structure of inorganic solids or metals. Passage of liquids and gases through solids leads to deterioration of most engineering materials. Effusion of gases through polymers was formulated by Graham (Mason and Kronstodt (1967) and is defined by:

$$\left(\frac{Rate_1}{Rate_2} \right) = \left(\frac{M_2}{M_1} \right)^{0.5} \tag{3.2}$$

Rate is defined as the rate of velocity of gas or liquid and *M* is the molar masses of gases 1 and 2 respectively. The diffusion/absorption/desorption of gases, vapours or liquids through porous polymers can also be explained by different mechanisms (Karge and Weitkamp 2008), including continuum theory (Fick's First and Second Laws), steady state and non-steady state diffusion and diffusion via external driving forces, including electrical, thermodynamic and chemical composition gradients (Mehrer, 2007).

Moisture

Humid ageing is also applicable to plastics. There are two main categories of humid ageing. Firstly, there are physical processes, mainly linked to the stress state induced by matrix swelling and sometimes matrix plasticisation. This kind of ageing can occur in matrices of relatively high hydrophilicity (affinity with water). Highly cross-linked amine cured epoxies are typical examples of this behaviour. The second category of humid ageing involves a chemical reaction (hydrolysis) between the material and water. Unsaturated polyesters are typical examples of this category. They display a low to moderate hydrophilicity – swelling and plasticisation have minor effects – but hydrolysis induces a deep polymer embrittlement and, eventually, osmotic cracking. Nevertheless, resilient fiber reinforced plastic (FRP) can resist water and display minor water concentration and its distribution in the sample thickness.

Ultraviolet light

UV light absorption has sufficient energy to cause chemical change to polymers, depending on the formulation of the polymer and the intensity and duration of exposure. Light that is absorbed is the key aspect of degradation and formulations, including UV absorbers, UV stabilisers and antioxidants, and inorganic screeners can provide better durability.

Biological interaction

Timber

The range of biological species that deteriorate and soften timber consist of fungi, beetles, termites and molluscs.

Fungi

Fungi are minute organisms that slowly digest the timber cell walls, leading to softening and decay. Fungi consist of an interweaved mass of spores and filaments. The filaments feed and eventually destroy timber, thriving within timber with moisture content greater than 20 percent and temperatures 20°C–40°C. 'Wet Rot' is a severe type of fungal attack, thriving within water-saturated timber and feeding on cellulose and lignin, resulting in a fibrous degradation along the grain of the timber. 'Dry Rot' is a fungus that is characterised by direct contact by timber with wet bricks or concrete within poorly ventilated spaces. The colour of the deteriorated timber is dark brown with cracks parallel and perpendicular to the timber (cubed), a result of the fungi consuming the cellulose. Decay seldom penetrates the heartwood because it contains fewer digestible constituents than the outer sapwood.

Beetles

Most types of beetles cause superficial damage to timber (Lyctid and Anobiid beetles); however, the Cerambycid ('Longhorn') beetles can destroy softwood structural timbers, although mostly the sapwood rather than the hardwood (structural) of the timber.

Termites

There are three types of termites that can lead to timber decay: subterranean, dampwood and drywood termites. Subterranean termites are sensitive to sunlight, and therefore they nest in large colonies underground, making mud tunnels to connect with the damp timber food source.

Termites tunnel along the length of the inner grain while feeding on cellulose and sugars and starches (all carbohydrates) – the insidious aspect of subterranean termites is that they provide little visible evidence of attack at the outside surface because the exposed timber surface remains intact while hollowing out the heartwood. Dampwood termites are larger than subterranean termites, but don't nest in the ground, nor do they leave visible signs in the ground (mud tubes) or holes in attacked wood. Dampwood termites tend to attack the sapwood and so reduce the risk of structural damage. Drywood termites do not require a source of moisture and, given moderate to high humidity conditions and warm temperatures, will infest dry, sound timber. The termites enter cracks and openings in wood to start new colonies. Wood is damaged as the drywood worker termites tunnel to enlarge their colony.

Teredo

Teredo molluscs, nicknamed 'Shipworm' because of the damage caused to medieval timber ships, attack underwater saturated timber structures such as seawalls and piles. Teredo, microscopic in size as larvae, enter saltwater-saturated timber and feed on the heartwood, gradually leading to hollow tunnels along the direction of the grain. Teredo can grow up to several metres in length; however, the damage is unseen from the outer surface of the timber until the damage can significantly affect the load-carrying capacity.

Crustaceans

Limnoria (or "Gribble") and Sphaeroma are crustaceans that congregate at the surface of exposed marine timber, and bore at right angles to the timber surface. Timber attack typically occurs on piles located within the tidal zone. In severe cases, the attacked timber pile can result in an "hourglass" appearance.

Sulphate-reducing bacteria and sulphate-oxidising bacteria

Concrete, metals and other engineering materials are vulnerable to microbial attack. Sulphate-reducing bacteria (SRB) are present underwater or underground and under anaerobic conditions generate hydrogen sulphide (H_2S). If the H_2S is aerated (e.g. construction excavation or within a sewer above the waterline), in the presence of sulphate-oxidising bacteria (SOB) and oxygen, the H_2S is transformed to corrosive sulphuric acid that deteriorates the cement binder within the concrete, leading to severe degradation and weakening of the cement binder (Figure 3.4).

Biofilms

Combined with electrochemical corrosion mechanisms, biofilms are congregations of microorganisms that grow on metallic surfaces, either above or below water, and lead to microbially influenced corrosion (MIC). The main types of bacteria that colonise metallic surfaces include SRB, SOB, iron oxidising/reducing bacteria, manganese-oxidising bacteria,

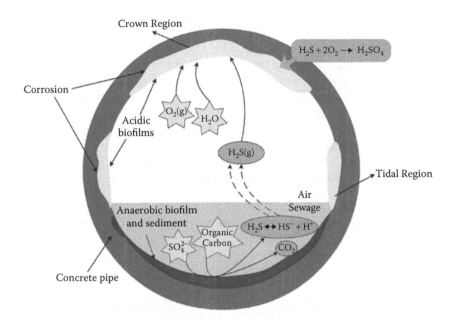

Figure 3.4 Mechanisms leading to biochemical generation of acidic exposure conditions within a sewer. (From Li, X. et al., *Front. Microbiol.*, 8, 683, 2017.)

bacteria-secreting organic acids and slime (Beech and Sunner, 2004). The anodic pH and chemical composition at the biofilm-metal interface is very different to the surrounding surfaces without biofilms (cathodic), and therefore corrosion becomes active. All MIC bacteria require metal ions for their growth, an energy source (e.g. UV light), a carbon source (e.g. CO_2 and organic materials), an electron donor that is oxidised and an electron acceptor (e.g. oxygen, SO_4^{2-} and nitrous oxides). The availability and the type of exposed metal have an effect on the colonisation of a metal surface.

Structural degradation

This mechanism overlaps with the *Physical Interaction* and *Mechanical* mechanisms discussed elsewhere in this chapter. Nevertheless, the daily utilisation of infrastructure leads to progress of ageing:

1. Overloading leads to cracking of reinforced concrete, excessive deflections and creep of timber, steel and fibre-reinforced composites, which, in combination with other ageing mechanisms, can lead to accelerated deterioration.
2. Settlement involves the downward movement of vertically loaded structural components, accompanied by consolidation of the

underlying soil. If the settlement is unequal, then the differential set-tlement can lead to cracking and excessive deflections.

3. Cyclic loading, for example a bridge, can lead to fatigue cracking, which will be discussed later in this chapter.

4. Differential movement and restrained expansion/contraction are sim-ilar to settlement, except the effects can be in three dimensions (rather than just vertical).

5. Brick growth involves the expansion of brick masonry with time, lead-ing to wall cracking. The amount of growth depends on the composition of the brick, the length of time between manufacturing and laying the bricks, and the amount of exposure to moisture and temperature ranges.

Most of the *structural degradation* aspects are accounted for by good design and workmanship, but nevertheless must be considered in combina-tion with other ageing mechanisms.

Chemical

There are overlaps between *chemical* deterioration to both *biological* and *electrochemical* ageing mechanisms, as well as the relationship of *physi-cal interaction* and chemical ageing. Repetition will be avoided in this section.

Hydrolysis

Hydrolysis is characterised by the chemical deterioration of a material due to reaction with water or deterioration due to absorption of as either free or bound water. It relates primarily to organic polymeric or polymer com-posite (e.g. fibre-resin composites) components of infrastructure. Exposure to moisture can occur underground (e.g. geosynthetic mats, commonly composed of polyester, polypropylene or polyethylene are buried to sta-bilise soils); waterproofing membranes to roofing and fascia components of buildings. The dual exposure to sunlight, or other aggressive chemi-cals (e.g. acids or alkalis), or elevated temperature (refer to T_g, glass transi-tion temperature, described earlier) can accelerate hydrolysis. Hydrolysis is accompanied by loss of stiffness and increased creep leading to premature failure. This process is typically slow and is largely affected by the equiva-lent diffusion coefficient of the composite material, which strongly depends on the type of reinforcing fibre (glass vs. carbon), type of resin (e.g. epoxy versus vinyl ester), sizing, layups implemented and manufacturing process.

Oxidation

This section specifically deals with the deterioration of engineering materials following reaction with oxygen (but specifically excludes electrochemical cor-rosion of metals, which is discussed in the electrochemical corrosion section (below)). Most plastics react slowly with oxygen, although when combined with elevated temperatures and UV exposure, the accelerated breakdown of the bonds within the polymeric chain can occur. Oxygen present within air-borne smog reacts with polymers with tertiary hydrogen atoms, such as polypropylene, rubbers and other polymers. The oxidised materials are inflexible, brittle and hence cracking occurs when they are stretched. Formulations that include antioxidants can provide better durability.

Acids

Exposure to acids can occur within a range of conditions, including airborne acidic pollutants and acid rain, ground and groundwater and industrial processes (refer to Chapter 4). Airborne acidic pollutants react with a range of materials and infrastructure, including concrete (depletion of the protective alkaline cover to reinforcement), stone masonry and metallic corrosion (to be discussed in the Electrochemical Mechanisms section). Acidic ground can chemically attack concrete by reacting with the cementitious binder; the speed of degradation is influenced by the concentrations of acids, extent of exposure (e.g. volume/exposed area ratio) and how mobile the groundwater is to replenish with fresh acid and remove reacted binder acid. Chemical resistance is good in Kelvar fibres, except for a few strong acids and alkalis (e.g. H_2SO_4 and NaOH) due to amide degradation. The range of industrial exposures to acids is vast and could be exhibited as blistering, cracking, embrittlement of polymer-based materials, reaction with the cementitious portion of concrete resulting in degradation of concrete and metallic corrosion (to be discussed later in this chapter). PVC is generally durable except in cases of exposure to concentrated acids (sulphuric acid above 70 percent concentration, hydrochloric acid (above 25 percent, and nitric acid at concentrations above 20 percent).

Alkalis

Concrete is alkaline with a pH of approximately 12.5 and therefore benign to alkali exposure. Nevertheless, particular aggregates within the concrete matrix can be deleteriously reactive under high alkaline conditions, and therefore vigilance with selection of raw materials is important. Depending on the type of metal and whether it is amphoteric (i.e. able to react both as a base and as an acid such as aluminium), the presence of alkaline conditions can create corrosive conditions. Polymers are mostly resilient to alkaline conditions; however, fibre composite (e.g. glass or kevlar) can be prone to hydrolytic deterioration in alkaline conditions.

Ozone

Ozone is created by the reaction of oxygen with ultraviolet light in the earth's atmosphere as well as electrical discharges. Chain breaking and cross-linking can both occur in polymers exposed to ozone, leading to cracking and embrittlement of rubber as well as deterioration of cellulosic and starch-based materials.

Solvents, fuels, alcohols, ketones, esters and aromatics

Each of these chemicals will need assessment on a specific case by case basis. Generally, concrete will be benign to these exposures, metals will need specific assessment of corrosivity, while polymers (e.g. FRP, sealants, and coatings) react differently depending on the formulation and can provide skin protection to exposed structural components.

Mechanical

The related aspects are discussed earlier in this chapter, namely abrasion.

Several of the factors described in *Physical Interaction* and *Structural Degradation* mechanisms (discussed earlier) overlap with *Mechanical* degradation and therefore duplication of particular factors will be avoided. The performance demands of built infrastructure necessitate that mechanical actions be accommodated. The worst-case scenarios are the resultant deterioration and loss of load-carrying capacity/functionality and potential collapse. The types of mechanical ageing are broad:

Surface wear due to abrasion

Erosion and cavitation cause the progressive weakening of an exposed surface and the gradual loss of section of the construction material. The degree of damage depends on the surface characteristics of the material (e.g. carburising a steel surface to provide additional hardness), amount of surface area exposed to wearing and friction effects influenced by the surface texture and roughness and whether protective treatments are applied (e.g. protective coatings).

Cavitation

In conditions where rapidly flowing fluids create local pressure below the vapour pressure, the situation is created for nucleation; a vapour bubble can form and will continue to grow until it moves to a region of higher pressure where it collapses. On spillways on concrete dams this can lead to gouging of the cementation binder, leading to localised erosion of the surface.

Dynamic actions

The range of dynamic actions imposed on infrastructure can include effects of impact, explosion, fatigue and vibration. The purpose of this book is to review ageing effects on infrastructure rather than the effects of extraordinary events arising from natural and man-made disasters, and therefore this section limits the discussion to fatigue and vibration effects. Examples of fatigue loading can include machine vibration, sea waves, wind action (e.g. wind turbines) and automobile traffic. Fatigue is defined as a process of cycle-by-cycle accumulation of damage in a material undergoing fluctuating stresses and strains (e.g. railway infrastructure). Corrosion fatigue relates to the joint mechanical degradation of a metal under the joint action of corrosion and cyclic loading.

Shrinkage/swelling

Porous materials such as concrete have a capillary pore network that changes in volume due to movement of water into the concrete (swelling) or loss of moisture due to drying (shrinkage). Particularly when drying, and if restrained (e.g. a wall restrained along the base), tensile stresses are set up and potential cracking can occur.

Environmental stress cracking

Environmental stress cracking (ESC) is characterised by the degradation of a plastic resin by a chemical agent while under stress. Loading allows chemical solvents to enter the plastic via surface crazing and cracking, thereby adversely affecting the polymer chains. The more rapidly that the chemical agent is absorbed, the faster the polymer will be subjected to crazing and subsequent failure. The types of failure are typically brittle fracture, evidence of multiple cracks and smooth fracture surfaces. This is common with amorphous plastics with low molecular weight and lower crystallinity while subjected to tensile stress, and the types of aggressive chemical agents include organic esters, ketones, aldehydes, aromatic hydrocarbons and chlorinated hydrocarbons.

Creep

Creep relates to the irreversible deformation of materials when subjected to load over a period of time. It seldom affects metals at room temperature; however, it is a consideration for RC because it could lead to differential movements around the built structure that could lead to cracking and mismatches of fixtures such as windows. Elevated temperatures lead to faster rates and total amount of creep. Creep is common within plastic materials at room temperature, which deform continuously under load. Excessive creep leads to intolerable dimension changes, leading to creep rupture and failure.

Electrochemical corrosion

'If gold rusts, what then can iron do?' – Geoffrey Chaucer, *The Canterbury Tales*

Electrochemical corrosion has been recorded since ancient times. It can be considered as the reverse of manufacturing: for example, iron oxides undergo energy-intensive processing to produce steel, whereas corrosion converts the steel into more thermodynamically stable iron oxides.

Corrosion is the deterioration of a metal, exhibited by loss of mass, caused by, for example, coupling of two different metals by touching (electronic) or electrochemical (connected by a conductive electrolyte). It also occurs on one metal surface where there are surface heterogeneities coupled with adjacent more uniform surfaces (cathodic) and connected by a conductive electrolyte. 'Rust' specifically is the corrosion product(s) of ferrous (iron) whereas 'corrosion' refers to the oxidation reactions of other metals.

Corrosion involves the deterioration of a material (anode) due to interaction with a different material or the same metal due to different exposures (cathode) where there is an electrical potential difference facilitated by a conductive path (electrolyte).

The anode loses electrons, for example, the anodic oxidation of iron:

$$Fe_{solid} \rightarrow Fe^{2+}_{aqueous} + 2e^{-} \tag{3.3}$$

This must be accompanied and balanced by a cathodic reaction that accepts the electrons and reacts with the dissolved iron. For example:

$$O_2 + 2H_2O + 4e^{-} \rightarrow 4OH^{-} \tag{3.4}$$

The balanced dissolution of iron can be represented by Equation 3.5, which couples the earlier two equations:

$$Fe^{2+} + 2OH^{-} \rightarrow Fe(OH)_2 \tag{3.5}$$

Depending on the availability of oxygen and water and acidity, Equation (3.5) may further transform into various types of iron oxides/hydroxides. For Equation 3.5 to proceed, there needs to be an electrical pathway (either a bridging electrolyte, such as water containing dissolved ions or sufficient humidity and temperatures to precipitate dew) between the anode (Fe_{solid}) and cathode (H^+). This is a very specific example because a wide variety of actions could lead to iron corrosion, and therefore a variety of oxides and hydroxides can be the resultant products.

As well as steel, awareness is needed of the corrosion properties of other metals and nonmetals used to construct infrastructure, including stainless steel (primarily an alloy of iron, nickel and chromium), galvanised steel, aluminium, copper, bronze, titanium and other construction alloys.

Each metal has a different electrical reactivity, which is evidenced as a standard redox potential, the electrical potential that indicates the readiness of the metal to receive or donate electrons (particularly important when considering contact between two dissimilar metals). The redox potentials are tabulated into an electrochemical series (refer to *NACE Corrosion Engineers Handbook*, pages 127–133; Baboian, 2016). When in either direct electrical contact or electrochemical contact via a conductive path, the anodic metal corrodes and donates electrons to a metal with higher redox potential that accepts the electrons from the anodic metal.

A comprehensive treatment on the subject of corrosion is provided in Ciek (2014) and Uhlig's Corrosion Handbook (Revie, 2011).

Atmospheric corrosion

Air-exposed metals tend to grow a thin protective oxide film on the surface (Brimblecombe, 2003). The types of exposures vary enormously and can include marine, industrial, tropical, arctic, urban and rural (refer to Chapter 4). The corrosion of metals in the airborne environment needs a protagonist to change the passive surface conditions (e.g. wind-blown chlorides in a marine environment and/or deposition of acidic pollutants, such as carbon dioxide, sulphurous and nitrous oxides). Once initiated and to support corrosion, the corroding anode needs an electrochemical pathway to adjacent passive steel (cathode) and this often includes surface moisture, rain, fog and humidity condensation due to temperature changes (dew). Wet deposition of 'acidic rain' containing SO_2 and NO_2 is also a source of corrosion initiation of exposed metals. Baboian (2016) provides corrosivity data for a broad range of geographical and climatic exposures. Exposure conditions vary with geometry of the exposed metal due to different exposures to sunlight, prevailing winds, amount of shade and surface temperature. Protection can be provided by galvanising and/or protective coating systems.

Galvanic corrosion

As discussed earlier, the contact between two dissimilar materials of different redox potential, whether directly in contact (electronic) or via an electrolyte (electrochemical), leads to galvanic corrosion. Crucial influencing factors include the conductivity of the electrolyte, the magnitude difference between the redox potentials of the two metals, combined with the ratio of exposed surface area of anode relative to cathode (with low ratio leading to faster rates of corrosion and leading to corrosion pitting).

Note that nonmetals can participate in galvanic corrosion, as exhibited by carbon that acts as a noble metal, lying between platinum and titanium in the redox series. Carbon fibres should not be used where they can come into contact with structural metals (especially light alloys where the carbon fibres are cathodic, such as aluminium or magnesium) in the presence of a conducting fluid (e.g. seawater).

Pitting and crevice corrosion

Pitting corrosion occurs when the anode is very small relative to the cathodic area, causing localised and accelerated corrosion. Pitting can be caused by local profile inhomogeneities on the metal surface, local loss of passivity, mechanical or chemical rupture of a protective coating/film, galvanic corrosion (described earlier) or the formation of a metal ion or oxygen concentration cell under a solid deposit (crevice corrosion).

Crevice corrosion is related to pitting; however, it is a type of attack that occurs within narrow openings between metal-to-metal or nonmetal-to-metal components. This environment has significantly lower oxygen content than the adjacent environment, forming a concentration cell with the adjacent oxygenated metal. This leads to low pH within the crevice thereby accelerating the rate of corrosion.

Concentration cells

When a portion(s) of a metal is exposed to electrolytes of different concentration, there is a concentration difference that promotes a difference in redox potential. When connected, the metal residing in the lower concentration becomes the anode. If the difference in potential is great enough, the more anodic area corrodes preferentially by concentration-cell corrosion. Similarly, electrolytes with different concentrations of oxygen, or differential aeration cells, will be corrosive, with the anode formation at the region of low oxygen concentration.

pH

The pH of the electrolyte exacerbates the rate of oxidation of an exposed metal, with preference to neutral or basic pH environment for durability. The effect of pH and redox potentials on corrosivity of particular metals are summarised as Pourbaix Diagrams, subdivided into 'Immune',

'Passive' and 'Corrosive' zones within a large selection of metals. Pourbaix Diagrams for a broad range of metals and exposures are provided in the *NACE Corrosion Engineers Handbook* (Baboian, 2016).

Corrosion combined with physical processes

Corrosion is exacerbated by simultaneous physical processes, including fatigue, stress, erosion and cavitation. *Corrosion fatigue* occurs when there is cyclic or fluctuating loading of a metallic structure (transportation, industrial, aerospace or defence) infrastructure that is simultaneously exposed to corrosives. The highly stressed regions undergo enhanced fatigue, evidenced by cracking, and undergo a more premature failure (than, say a metal subjected to fatigue conditions). A comprehensive coverage of corrosion fatigue can by sourced from Crooker and Leis (1983) that approaches the mechanisms, experimental observations and engineering aspects of corrosion fatigue. *Stress corrosion cracking* is the result of combined tensile stress, applied either in service or residual stresses caused by manufacturing (e.g. welding) within a component that is also exposed to a corrosive environment. It is accompanied by tensile cracks and early failure. The theory, methods of testing and case studies are provided in Raja and Shoji (2011). *Erosion corrosion* is characterised by exposure of a metal to a corrosive flowing liquid, resulting in combined physical wear and corrosion. This can occur within a broad range of infrastructure types including energy, water and wastewater, and industrial applications. The increased turbulence caused by pitting on the internal surfaces, the turbulent flow conditions, and the composition and velocity of the fluid can result in rapidly increasing erosion rates and eventually a leak. Matsumura (2012) provides a detailed coverage of the history, mechanisms and theory of erosion corrosion as well as observations from field studies. *Cavitation corrosion* occurs where there is flowing liquid in contact with a metal that is subject to a change in pressure, accompanied by implosion of gas vapour bubbles at the metal surface, causing localised pitting corrosion. This can occur within metallic pipelines, vessels, pumps, water turbines and at locations where geometry changes (e.g. pipe elbow). Cavitation corrosion affects a range of built infrastructure types, including the industrial, energy, maritime, water and wastewater industries. This mechanism can be accompanied by erosion corrosion in adjacent areas. Franc and Michal (2004) provide a comprehensive treatment on cavitation corrosion.

Microbial corrosion

This is a vast subject that has been partially covered earlier in this chapter regarding deterioration of plastics, timber and concrete by microbial organisms. MIC conveys the deterioration of metallic materials by organisms. The range of bacterial corrosion problems can be attributed to the presence of sulphate-reducing bacteria (SRB) and other types of bacteria, such as iron-oxidising bacteria, slime-forming bacteria and acid-producing bacteria, and in some cases other organisms, such as sulphur bacteria, moulds, yeasts and algae. Most microorganisms attach to solids, colonise and form biofilms that produce an environment at the biofilm/metal interface that is radically different from the bulk medium in terms of pH, dissolved oxygen, organic and inorganic species.

The most important bacteria in the corrosion process are those involved in the sulphur cycle. These include two groups: sulphate-oxidising bacteria (SOB) and those involved in the reduction of sulphur, notably sulphate-reducing bacteria (SRB). SRB are the most significant bacteria, prolific in the oil and gas industries, accelerating the corrosion of iron by removing atomic hydrogen from the iron surface through the bacterial enzyme hydrogenase. The removed hydrogen reacts with sulphide produced by the SRB, forming H_2S, which is known to be a toxic and corrosive gas.

Most MIC, however, manifests as localised corrosion because most organisms do not form in a continuous and uniform film on the metal surface.

Slime-forming bacteria are mostly aerobic and produce corrosive organic acids. These bacteria exist more frequently in fresh rather than saline water, commonly causing corrosion in the oil industry (pipelines and equipment, inject wells and storage tanks).

Iron-oxidising bacteria derive energy from the oxidation of ferrous (Fe^{2+}) to ferric (Fe^{3+}) hydroxide at or near neutral pH, and in some cases, the result is the formation of dense deposits of iron oxides. It can set up oxygen concentration cells that promote corrosion and can establish an anaerobic environment.

A more detailed coverage of MIC can be sourced from Loto (2017).

High temperature corrosion

Temperature affects the kinetics of corrosion, with faster rates expected at higher temperatures. At extremely high temperatures, metals are oxidised in the form of surface scaling. If the formed oxide scale is thin, slow-growing and adherent, it protects the substrate from further oxidation. However, if the scale spalls frequently, the metal is consumed continuously and the material ultimately fails. Further detail is available from Khanna (2013) and Lai (2006).

Stray-current and interference corrosion

Stray current is characterised by electrical current that strays from the intended electrical pathway. Within DC-electrified railway systems, as well as providing navigation, the rails provide the returning path of the train's electric current to the traction power substation (TPS). Due to electrical resistance of the rails and ground conductivity, and because the rails are not totally earth-isolated, a part of the returning current to the TPS strays into the ground.

Stray electrical currents can be transmitted through the soil and migrate to nearby buried structures at one point only to return to the soil at another point, leading to corrosion at the point of exit. The amount of built underground infrastructure is vast, comprising pipelines, cables, tunnels, storage tanks, basements and foundations. The electrolytic corrosion occurs at the point that stray current flows out of the metallic structure where an anodic corrosion reaction takes place, leading to oxidation of the metal. Other sources of stray current corrosion occur when cathodic protection systems, installed for corrosion control of particular underground/underwater facilities (e.g. pipelines), leak stray DC current to nearby unprotected facilities, electroplating units and DC-braking systems on light rail. Solutions to the problem include reasonable spacing of traction substations, continuously welded track rails, high track-to-earth resistance or isolation, bonding rails to the elevated structure to further reduce the resistance of the negative return circuit and bonding of the rail return current path to a parallel water main or pipeline (the bond providing a metallic path for the current to follow rather than discharging off the rail).

Corrosion of steel in reinforced concrete

The high pH of concrete provides passive conditions for the embedded steel, which is protected by a thin iron oxide film. If these conditions remain unchanged, then the steel is durable. If the conditions of high pH or the passive iron oxide change, the steel actively corrodes.

Concrete has a documented ability to chemically react with carbon dioxide and other acidic airborne gases, referred to as carbonation. Gaseous CO_2 penetrates into concrete by diffusion through unsaturated concrete pores and reacts mostly with calcium hydroxide. During the Service Life of a built concrete structure, concrete progressively carbonates with increasing depth. When the depth of carbonation, characterised by depletion of alkalinity, reaches the depth of the steel reinforcement, this leads to the breakdown of the protective iron oxide film at the steel surface and corrosion commences.

An alternative means of corrosion initiation occurs when chlorides are present. This can occur where there is exposure to marine conditions or when chloride is in contact (e.g. exposure of bridge decks to chloride-based deicing salts) or cast into fresh concrete (e.g. chloride-based set accelerator) (Figure 3.5).

A widely used conceptual model for Service Life prediction of corroded RC structures idealises a two-stage corrosion process: an initiation stage, during which carbonation/chloride ions reach the reinforcing steel

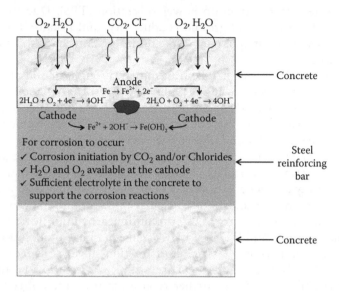

Figure 3.5 Schematic diagram of reinforcing steel corrosion in concrete as an electrochemical process. (From Ahmad, S., Figure 3.1 from Service Life *Cem. Concr. Compos.*, 25, 459–471, 2003.)

in sufficient quantities to depassivate the steel and a propagation stage, in which expansive corrosion takes place until an unacceptable level of induced cracking of the concrete cover has occurred (Figure 3.6). Following corrosion initiation, the resultant iron oxides induce volumetric expansion of the steel bar. Depending on the oxidation state, iron can expand as much as six times its original volume. The volumetric expansion exerts tensile stresses in the concrete surrounding the reinforcement, and when these stresses exceed the concrete tensile strength, cracking of the concrete cover occurs.

Firstly, the initiation period involves (insert) transport of the corrosives (and delete) to the transported though the cover concrete to the steel. Following corrosion initiation, the kinetics of corrosion is critical, either slowing the accelerating corrosion rate. The iron oxides/hydroxides are magnitudes larger than the passive iron oxide (prior to initiation), thereby exerting tensile pressures on the cover concrete and resulting in eventual cracking and spalling.

When a RC structural member is loaded, transmission of force relies on the capacity of the bond stress developed between the concrete and steel along the length of the bar. Corrosion can significantly affect the rebar-to-concrete bond strength; the bond strength mechanism comes partly from the frictional contact forces that develop along the bar-concrete interface,

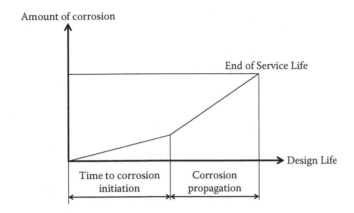

Figure 3.6 Common 2-stage steel reinforcement corrosion damage model over time. (Courtesy of F. Collins.)

and partly from interlocking of the bar ribs and the surrounding concrete. Bond can be sustained as long as the radial splitting components are supported. However, once an external longitudinal crack has formed along the bar, the confinement provided by the cover concrete is reduced, as is the frictional component of bond, accompanied by a significant reduction of stiffness and strength for the overall structural member. As cracks propagate, the cover concrete finally detaches.

Because this is a very broad subject, more detailed and specific information can be sourced from Broomfield (2006) and Bertolini et al. (2013).

CONCLUSION

Identification of the types of deterioration related to the subcomponents of built infrastructure is critical in the determination of the mechanisms leading to that deterioration. The mechanisms are many and varied — it is critical that the exposure conditions, geometry, material type and scale of the built component be examined in a case-by-case manner. This chapter has reviewed the mechanisms of ageing and provided the reader with relevant sources of background references, which provide greater detail.

REFERENCES

Ahmad, S. (2003). Reinforcement corrosion in concrete structures, its monitoring and service life prediction—A review, *Cement and Concrete Composites*, 25(4–5): 459–471.

Baboian, R. (2016). *NACE Corrosion Engineer's Reference Guide*, 4th ed. Houston, TX: National Association of Corrosion Engineers, 568p.

Beech, I. B. and Sunner, J. (2004). Biocorrosion: Towards understanding interactions between biofilms and metals, *Current Opinion in Biotechnology*, 15: 181–186.

Bertolini, L., Elsener, B., Pedeferri, P., Redaelli, E., and Polder, R. B. (2013). *Corrosion of Steel in Concrete: Prevention, Diagnosis, Repair*, 2nd ed. Weinheim, Germany: Wiley-VCH, 434p.

Brimblecombe, P. (2003). *The Effects of Air Pollution on the Built Environment*, London, UK: Imperial College Press, 422p.

Broomfield, J. P. (2006). *Corrosion of Steel in Concrete: Understanding, Investigation and Repair*, 2nd ed. London, UK: Taylor & Francis Group, 296p.

Burks, B. and Kumosa, M. (2012). The effects of atmospheric aging on a hybrid polymer matrix composite, *Composites Science and Technology*, 72(15): 1803–1111.

Cicek, V. (2014). *Corrosion Engineering*. Beverly, MA: Wiley, 288pp.

Collins, F. G. and Sanjayan, J. G. (2008). Unsaturated capillary flow within alkali-activated slag concrete, *Journal of Materials in Civil Engineering*, 20(9): 565–570.

Collins, F. G. and Sanjayan, J. G. (2009). Prediction of convective transport within unsaturated concrete utilizing pore size distribution data, *Journal of Porous Materials*, 16: 651–656.

Crooker, T. W. and Leis, B. N. (1983). *Corrosion Fatigue: Mechanics, Metallurgy, Electrochemistry, and Engineering, STP801*. Philadelphia, PA: ASTM International, 531p.

Franc, J. P. and Michal, J. M. (2004). Fundamentals of cavitation, *Fluid Mechanics and its Applications*, Volume 76, Dordrecht, the Netherlands: Kluwer Academic Publishers, 306p.

Hall, C. and Tse, T. K. M. (1986). Water movement in porous building materials—VII. The sorptivity of mortars, *Building and Environment*, 21(2): 113–118.

Karge, G. and Weitkamp, J. (Eds.). (2008). *Adsorption and Diffusion*. Berlin, Germany: Springer, 400p.

Khanna, A. S. (2013). High temperature oxidation. In M. Kutz (Ed.), Chapter 5 of *Handbook of Environmental Degradation of Materials*, 2nd ed. Burlington, NJ: Elsevier, pp. 105–152.

Lai, G. Y. (2006). *High Temperature Corrosion and Materials Applications*. Materials Park, OH: ASM International.

Law, F. L. and Evans, J. (2013). Effect of leaching on pH of surrounding water, *ACI Materials Journal*, 110(3): 291–296.

Li, X., Kappler, U., Jiang, G., and Bond, P. L. (2017). The ecology of acidophilic microorganisms in the corroding concrete sewer environment, *Frontiers in Microbiology*, 8: 683.

Loto, C. A. (2017). Microbiological corrosion: Mechanism, control and impact—A review, *The International Journal of Advanced Manufacturing Technology*, 92(9–12): 4241–4252.

Lucija, H., Kosec, L., and Ivan, A. (2010). Capillary absorption in concrete and the Lucas–Washburn equation, *Cement & Concrete Composites*, 32: 84–91.

Mason, E. A. and Kronstodt, B. (1967). Graham's Laws of diffusion and effusion, *Journal of Chemical Education*, 44(12): 740–744.

Matsumura, M. (2012). *Erosion-corrosion: An Introduction to Flow Induced Macro-cell Corrosion*. Oak Park, IL: Bentham Science Publishers, 160p.

Mehrer, H. (2007). *Diffusion of Solids: Fundamentals, Methods, Materials, Diffusion-Controlled Processes*, Springer series in solid state science 155. Berlin, Germany: Springer-Verlag, 639p.

National Transportation Safety Board. (2008). Collapse of I-35W Highway Bridge, Minneapolis, Minnesota, August 1, 2007. Highway Accident Report NTSB/HAR-08/03.

Raja, V. S. and Shoji, T. (2011). *Stress Corrosion Cracking: Theory and Practice*. Woodhead Publishing, 816p.

Revie, R. W. (2011). *Uhlig's Corrosion Handbook*, 3rd ed. New York: John Wiley & Sons, 1296p.

Summerscales, J. (2014). Durability of composites in the marine environment. In: P. Davies and Y. D. S. Rajapakse (Eds.), *Durability of Composites in a Marine Environment, Solid Mechanics and Its Applications 208*. Dordrecht, the Netherlands: Springer Science+Business Media, pp. 1–13.

Washington, DC, 178p. https://www.ntsb.gov/investigations/AccidentReports/Reports/HAR0803.pdf.

Zhutovsky, S. and Hooton, R. D. (2017). Experimental study on physical sulfate salt attack, *Materials and Structures*, 50(54): 1–10.

Chapter 4

Environmental exposure

INTRODUCTION

Weather is arguably the most common subject of conversation (and complaint!) within our daily lives. Just as exposure to sunlight ages our skin, exposure to ultraviolet radiation degrades polymers by breaking molecular bonds; if an exposed elastomeric sealant on a building could talk, it would vainly complain of losing sheen and gloss over time, accompanied by cracks, wrinkles and chalking of the skin.

Built infrastructure coexists with an external environment that can have a diverse range of physical and chemical characteristics. It is critical that the habitat of built infrastructure is properly understood because even mildly corrosive conditions can significantly age construction materials over the term of the Service Life. Characterisation of the exposure environment, in the context of the mechanisms of ageing, as discussed in Chapter 3, is therefore an important step when assessing the ageing of built infrastructure. In broad terms, we can classify environmental exposure into the *macro-*, *meso-* and *micro-*environment types.

MACRO-ENVIRONMENT

The macro-environment affects large geographical regions and is applied to multiple built infrastructures that may be located in one or several countries. The macro-environment can be identified by common features that define the climate, terrain, vegetation, land use, geology, size of the land mass and proximity to watercourses and oceans. The macro-environment is difficult to change: the features that define the physical and human geography of a region are reasonably fixed. However, these macro-environmental characteristics must be understood when assessing the effects of ageing.

Characterising the macro-environment

For over a hundred years, the Köppen–Geiger System has been used by geographers to characterise global climate in terms of five major zones based on temperature and precipitation (Peel et al. 2007; Marsh and Kaufman 2012). However, regions are also affected by the amount of solar exposure, regional winds, cloud cover and terrain, and therefore reliable long-term climatic and geographic data is essential for the characterisation of the macro-environment.

A key source of regional climatic data is the World Meteorological Organization (WMO, 2015), which also provides useful links to the many national bureaus of meteorology. Land-based stations as well as satellite and radar measurements can be sourced as raw data or visualised regionally by interactive Geographical Information Systems (GIS)-based mapping tools. Relevant data includes the hourly, daily and monthly air temperature, relative humidity, precipitation, direction and speed of the prevailing winds, duration and intensity of exposure to solar radiation, intensity and frequency of snowfall and frost. Air pollution composition and concentrations (e.g. sulphur dioxide and potential for acid rain) can be gleaned from key international agencies, for example, the European Pollutant Emission Register (E-PRTR, 2015) and national pollutant inventories such as the Australian National Pollutant Inventory (NPI, 2015).

Utilisation of national weather models based on meteorological data for the purpose of quantifying the environmental 'loading' on ageing infrastructure is demonstrated by LifeCon, Life Cycle Management of Concrete Infrastructures for Improved Sustainability (Carlsson, 2003). LifeCon provides instructions and guidelines on how to characterise the macro-environmental loads onto infrastructure and provides example data and analysis of five European countries (Norway, Sweden, Germany, Finland and United Kingdom) toward the systematic evaluation of environmental loading toward the durability assessment of concrete urban, maritime and regional infrastructure. International codes of practice have followed, for example, ISO 9223, ISO 9224, and AS5604.

Regional geographical databases provide vital background to the terrain, geology, vegetation types and coverage, regional river and groundwater systems, oceans, types and concentrations of land use (e.g. urban/city, rural, industrial or mixed). Considerable data is available via two types of database: either a spatial warehouse that provides tabulated measurements or via GIS-based mapping tools that allow interactive analysis of the data. The International Union of Geological Sciences (IUGS, 2015); U.S. Geological Survey (USGS, 2015), and similar national authorities provide a wealth of spatial databases, including topographic relief over the region, underlying ground stratigraphy regarding soil and rock types, composition and flows of groundwater and watercourses, characteristics of oceans and shorelines and adjacent landmasses, land cover by vegetation and types of land use.

Proximity of oceans and large watercourses has a profound effect on regional climate: including higher precipitation and moderated air temperatures, lake- or ocean-affected snow from cold air moving across a warm body of water, strength and direction of prevailing wind. National and international oceanographic databases are key sources of information on the direction and speed of ocean currents, water temperatures and composition (e.g. corrosive salinity), the underlying profile of the sea bed and nearby land, tidal information and ocean climatology (IODE, 2015; NODC, 2015).

Comparison of two different macro-environments

The coastal plains on the Gulf Coast of Eastern Saudi Arabia have close proximity to an ocean with high salinity (Rasheeduzzafar et al., 1984; Haque et al., 2007). Low precipitation (<50 mm/year) combined with high evaporation (>1,200 mm/year) provide conditions where there is high salinity at ground level. Despite the arid conditions and low rainfall, the humid conditions at the coast are conducive to significant wetting of exposed surfaces by dew deposited during the evening. High exposure to solar radiation is complemented by consistently high temperatures that are generally greater than 30°C and sometimes exceeding 50°C. A shortcoming of the Köppen–Geiger classification is that the Gulf Coast would be classified as Bw (Desert), whereas the very corrosive saline and humid conditions would not be considered.

Engineering materials exposed to these conditions are highly vulnerable to deterioration. The highly saline ocean and soil and air provide very corrosive conditions to exposed metals, and this is exacerbated by regular surface wetting by dew and high ambient temperatures. Elevated temperatures and exposure to ultraviolet light cause embrittlement of polymers and plastics due to photolytic and photo-oxidative reactions. Exposed porous materials (e.g. masonry and concrete) are vulnerable to deterioration arising from expansive salts crystallising at the surface.

A stark contrast with the Arabian Gulf occurs in the cold extremes of northern Sweden, a region exposed to a climate characterised by long winters (mean temperatures rarely above 0°C and minimal solar UV) and short summers (peak temperature is seldom greater than 15°C). The topography is generally flat lowland although mountainous near the western border with Norway. In the colder climate, there will be different mechanisms of deterioration. Freeze-thaw cycles lead to surface abrasion and attrition of exposed infrastructure. When water freezes within exposed porous materials, there is accompanying expansion (approximately 9 percent) by ice crystals within the pores, leading to tensile pressures that cause disruptive cracking and a progression of failure at the surface. Ductility of many materials decreases at low temperatures, causing brittle failure of exposed metals, ceramics and polymeric materials (Boyd 1970; Myer 2013).

These two examples paint a background of two exposure extremes, each having considerable affect on the ageing of built infrastructure located in these regions.

MESO-ENVIRONMENT

While the macro-environment describes large geographical regions that include cross national boundaries, the meso-environmental regions are comprised of many localities that are exposed to unique meso-environments that influence ageing. The locality level can comprise an area with common physical features (such as a river valley) or man-made features such as a city.

Suppose two identical buildings share the same macro-environment; however, one of the buildings displays advanced deterioration. Why? In this example, the first building is located 50 metres from the coast. The meso-environment is corrosive with exposure to airborne chloride from the sea. The second building resides in a rural location that is 200 km inland and not exposed to corrosives. The following discussion will consider the types of meso-environments that affect ageing of infrastructure.

Airborne salts

Concentrations of chloride play a key role in the formation of anodes on metallic surfaces, leading to corrosion. Breaking sea waves eject microscopic droplets and aerosol of waterborne salt into the air (typically 0.1 to 1000 μm diameter) and are transported inland under the influence of prevailing winds, with concentrations diminishing within approximately 400 metres inland (Feliu et al., 1999). The intensity of deposition of airborne salts needs consideration and depends on many factors, including (Lewis and Schwartz, 2013):

1. Strength, frequency and direction of the prevailing winds.
2. Width and proximity to shore of the surf zone, intensity of the breaking waves and tidal variation and movements.
3. Wind–land interfaces that can deflect wind. The topography of the coastline and inland areas has a key influence, evidenced by lower transport inland of aerosol in areas of rugged terrain.
4. Proximity and type of vegetation (e.g. coastal forests can imbibe significant salt aerosol by foliar absorption).
5. Land use, whereby a city landscape with a mixture of built geometries, heights and sizes of infrastructure will deflect and circulate air with lower and more variable wind velocities than in rural or less built-up areas.
6. Shelter from prevailing winds provided by harbours or adjacent buildings.

Airborne pollutants

Background levels of acidic gases (CO_2, SO_2 and NO_2) in the air become more concentrated within industrial and city/urban locations due to higher amounts of emissions arising from burning of fossil fuels. At particular locations, the cooler air pollutants become trapped at ground level by an overlying layer of warm air, commonly termed a 'winter temperature inversion'. The effects of the inversion are made worse where the terrain and urban topography restricts the dispersion of the gases by prevailing winds.

Acid deposition is characterised by either dry deposition, where windborne gases are deposited on the surface, and/or wet deposition, where the gas reacts with water (in the form of rain, dew, water vapour or snow) forming acids (or 'acid rain').

Deterioration by exposure to airborne acids has been reported for European historic and modern stone masonry structures and monuments (Dolske, 1995; Sabbioni, 2003), progressing corrosion of exposed metals and diminution of the high alkalinity within concrete that eventually initiates corrosion of the embedded steel reinforcement.

Ultraviolet radiation

The macro-environment has already considered the geographical location as the first clue to exposure to solar radiation, as a function of latitude and daily/seasonal variations in the sun elevation. If the locality is at high altitude, the infrastructure will be exposed to greater UV radiation than a location at sea level due to the air atmosphere absorbing less radiation. Nearby reflective surfaces can exacerbate UV exposure; for example, snow and water have high albedo whereas adjacent forests absorb considerably more UV (Blumthaler and Ambach, 1988). Thick cloud coverage can attenuate UV exposure due to water droplets absorbing radiation; however, the filtering is minimal where the cloud coverage is thin. The locality may be shadowed by adjacent buildings or hilly terrain and therefore has lower exposure than unsheltered infrastructure. Depending on the function of the infrastructure and the geometric configuration of components (e.g. horizontal, inclined or vertical), shading and/or protective surface treatments can affect solar exposure.

It is the UVA spectrum with 315–400 nm and UVB with 280–315 nm wavelengths that are of most concern with the durability of engineering materials. Exposure to UV radiation affects most polymer-based materials, causing photo-oxidative reactions that degrade protective coatings and paints, membranes, sealants, plastics and polymer composites, although the effects can be modified by chemical composition, inclusion of UV stabilisers and protective gel coats (Chin, 2007).

Surface wetness

Within a meso-environment, exposure to rainfall can bring mixed fortunes: rainfall can contribute beneficial washing of corrosive salts from the surface, whereas pooled water and prolonged surface wetness can accelerate corrosion of unprotected metals. Atmospheric corrosion is an electrochemical process, relying on a sufficiently conductive medium that will facilitate the chemical reactions between the anode and cathode; therefore, a meso-environment that promotes surface moisture via regular precipitation or condensation (dew) will have higher 'time of wetness' and corrode at a faster rate.

Infrastructure exposed to regular wetting and drying may also be prone to higher accumulation of corrosive salts at the surface of porous materials (e.g. concrete, masonry and terra cotta). Prolonged wet/dry cycles and accumulation of high surface chloride can lead to long-term diffusion into porous materials; this renders embedded metals (e.g. steel in the case of reinforced concrete) vulnerable to future corrosion. A different problem occurs under wet/dry conditions where the expansive crystallisation of salts in the surface pores can crack and exfoliate porous materials. This can be exacerbated by drying conditions, including the interplay between exposure to winds, temperature and relative humidity.

Specific wetting exposures, whether industrial process liquors, groundwater, wastewater or maritime exposures should be treated as micro-environments.

Underground exposure

Considerable infrastructure lies beneath the land surface (e.g. tunnels), sea (e.g. submarine pipelines) or partially underground (e.g. foundations that support a building). The meso-environment must consider the soil and groundwater chemical composition, permeability and the propensity toward corrosion and/ or adverse chemical reaction with construction materials. Key data can be gleaned from local geological and underground surveys, records of groundwater and river systems, past land use and past construction records from within the locality. Within a locality, comprising many built infrastructure, the ground properties can vary considerably, and the project considerations of specific components of built infrastructure will be discussed in the following paragraphs within micro-environments. Good background to ground durability can be found in (Robery, 1988; Beavers, 2001; BRE, 2005; Doran et al., 1987).

The meso-environment contains key risks that may pose a deterioration risk:

Acidity

Soil acidity can occur either naturally or by man-made causes. Natural acidic soils are formed over long geological periods, as evidenced by chemical weathering of acid parent rock (e.g. granite) that release hydrogen anions. This can be exacerbated by a wet climate where there is significant leaching of base cations (calcium, potassium, magnesium and sodium) from the weathered soil. Decomposition of organic matter within the soil is another contributor to the formation of acids.

Acidity can be accelerated by agricultural practices, commonly by leaching of nitrate from ammonium nitrate fertilisers, and also high accumulations of organic matter in the soil. Industrial environments can lead to underground acidity due to, for example, acid cleaning. The generation of acidic soil can also occur from nearby coal or metalliferous mines where metal sulphides are oxidised, thereby, generating sulphuric acid. Acid rain, discussed in Section 4.3.2, precipitates primarily sulphuric and nitric acids that permeate into a soil, reacting with and leaching base cations that progressively create a low pH soil. Apart from acidity, fertilisers containing ammonium sulphate and ammonium nitrate have a known track record of reacting adversely with cementitious binders. Microbiologically induced acidity is a special case that is discussed separately in the following paragraphs.

Microbiological

There is a diverse range of microorganisms that can thrive within many environments, both aerobic and anaerobic. The most common types that should be considered as harmful to buried construction materials are sulphur and sulphate-oxidising bacteria (SOB) and sulphate-reducing bacteria (SRB) (Doran et al., 1987; Javaherdashti, 2011; Okabe et al., 2018). In the case of exposed steel, microbiologically influenced corrosion (MIC) occurs when particular bacteria congregate at the surface of a buried metal, forming a biofilm that changes the normally passive conditions, resulting in acceleration of anodic or cathodic corrosion reactions. A carbon source is needed to sustain the bacteria (e.g. dissolved carbon dioxide, organic substances or carbon within steel), nutrients from the soil and water at the metal surface.

The same types of bacteria play a more insidious role. A soil may lie dormant while bacteria is happily feasting on iron and sulphates from the soil and excreting iron sulphides. If kept underwater, these soils are not harmful. However, if the soil is exposed to air (due to excavation during construction, or drainage resulting in a lower water table or below the water table, or during extreme drought), oxidation of the pyritic materials generates acidic conditions. Acid sulphate soils commonly occur in coastal estuarine floodplains, although pyritic sources can also occur inland. The conditions require the presence of iron-rich sediments, sulphate, removal of reaction products (e.g. bicarbonate), sulphate-reducing bacteria and organic matter.

Similar mechanisms occur within sewer and wastewater tunnel/pipeline environments, where sulphate-reducing bacteria convert sulphurous compounds to hydrogen sulphide (H_2S). In partially filled pipelines and tunnels, the gaseous H_2S rises above the liquid sludge and collects overhead at the walls of the pipeline where, in the presence of oxygen, Thiobacillus bacteria metabolise sulphuric acid, which has a significant ageing effect on many engineering materials.

Oxygen content

Underground corrosion of exposed metals and reinforcement within concrete is a complex subject. The electrochemical nature of corrosion necessitates reactions at the metal surface involving oxidation, commonly accompanied by oxygen reduction, facilitated by suitably low electrical resistance of the ground/concrete to facilitate the reactions. Oxygen access commonly controls the corrosion rate: a more permeable soil will permit a greater rate of transport of oxygen; however, the location of groundwater within the soil strata is a key factor because a 'wet' soil is almost impermeable to oxygen. Within hollow reinforced concrete structures (e.g. within a tunnel), with hydrostatic pressure on one side only, the groundwater will flow from the saturated concrete to a dew point is reached where the rate of evaporation from the air-exposed face equals the rate of groundwater penetration from the saturated side. This sets up a corrosive condition known as "hollow leg corrosion, where there is a galvanic corrosion, or differential oxygen cell, between the oxygenated embedded steel with the saturated face with low oxygen content. Also, the dew point can act as a zone where aggressive salts (e.g. sulphates) can accumulate and react adversely by chemically attacking the adjacent concrete (Silver et al., 1999). Differential aeration of the metal, commonly represented by a vertical steel pile with different soil exposures to oxygen along the length, is exacerbated by stratification of the soil composition and/or partial immersion in groundwater – higher rates of corrosion can occur when there are larger aerated zones and smaller areas of oxygen-starved steel. Soil resistivity characterises the ability of the soil as an electrolyte to facilitate electrochemical corrosion between the anodic and cathodic regions of the metal, influenced primarily by composition (e.g. soluble salts) and moisture content.

Specific construction configurations can create corrosion macrocells between zones of high and low oxygen content and need to be addressed as microclimates when considering particular built elements. An example is the case within water-excluding structures (e.g. reinforced concrete tunnel). One exposed face has readily available access to oxygen whereas the soil-exposed face is oxygen starved (Doran, 1987).

Stray current corrosion

The ground can also transmit stray electrical currents that accelerate corrosion of buried metals. Stray currents can arise from direct current (DC) railway and transit systems or from high-voltage underground services, where the current is transmitted through a lower resistance conductive pathway through the ground. Installations such as buried pipelines or tunnels may collect the stray current (cathode) and transmit it at another location back to a substation (anode), resulting in corrosion. Knowledge of nearby buried services and installations within the locality is critical when considering potential corrosion risks.

Ground salinity

Ground salinity affects considerable areas of the world (Rengasamy, 2006) that are not necessarily located in coastal or offshore areas but have climatic conditions of low rainfall and high rates of evaporation. It can also be affected by human activity, such as urbanisation or agriculture, causing the height of the groundwater to rise; salts found naturally in rocks and soil are dissolved and move upward toward the soil surface. The resulting corrosive soil is rich with chlorides and sulphates, has low resistivity and is highly conducive to corrosion of metals and chemical attack of a suite of construction materials.

Sulphates

Degradation of concrete can arise from exposure to soils containing sulphates (BRE, 2005). Sulphates react with calcium hydroxide (formation of expansive calcium sulphate) and aluminates (formation of expansive ettringite) in the cement paste, leading to expansion and breakdown of the cement paste. The severity of the attack will be a function of the types and concentration of sulphate, cement composition (e.g. amount of tricalcium aluminate [C_3A] in the cement), and whether the concrete is sufficiently permeable to facilitate ingress of sulphate. Magnesium sulphate poses a greater deterioration risk, evidenced by substitution of calcium by magnesium within the calcium silicate hydrate binder, which transforms the binder into a very weak material (Siad et al., 2015). A different type of sulphate attack involving thaumasite formation occurs when the ground temperature is less than about 12°C, high levels of sulphate that are present in the adjoining soil and abundantly available moisture, composition of the concrete (availability of carbonates) and cementitious binder type and composition (Rahman and Bassuoni, 2014).

'Soft' water

Ironically, very pure water (e.g. distilled water) can be harmful due to leaching of calcium and hydroxyls out of the concrete binder, resulting in dissolution of the calcium-bearing cementitious phases. Certain soils harbour 'soft' groundwater, characterised by low concentrations of dissolved salts, specifically low calcium content that is harmful to concrete. This will be influenced by the permeability of the soil and the presence of groundwater that can remove reactants and replenish the site with fresh corrosives. Sometimes these groundwaters contain dissolved carbon dioxide, and when dissolved in the groundwater, carbonic acid (H_2CO_3) is formed, and this will react with any hardened cement or limestone aggregate to form a weak calcium bicarbonate that is dissolved into the groundwater.

Background to the ground meso-environment can be gleaned from published geology, soil and groundwater maps and records that summarise the past results of ground chemical testing and past construction in the locality. Codes of practice also provide guidance on ground types, pH, chemical composition and resistivity. Guidance is also provided via ground corrosivity maps that are discussed in Section 2.2.7.

Meso-environmental mapping

With the aid of Geographical Information Systems (GIS), regional test data and predictive models (Cole et al., 1999, 2003), many countries have assembled environmental and corrosivity maps for many localities. For example, air corrosivity maps of Australian regional centres (King and Moresby, 1982; Cole et al., 1999; Carlsson, 2003) and atmospheric corrosion in coastal regions within a range of countries (Slamova et al., 2012).

Ground mapping includes the soil corrosivity in the United Kingdom (BGS, 2015), ground salinity in coastal China (Guo et al., 2013), salinity maps of the European Union (Toth, 2008), acid sulphate soil in Australia (Naylor, 1998; Huang et al., 2014), global soil properties and land use, ecosystems and water resources (IGBP-DIS, 1998).

LifeCon Methodologies (Carlsson, 2003) have utilised computational fluid dynamics (CFD) that considers wind–flow effects on chloride deposition in an airborne coastal environment. The CFD comprises 3D models, the effects of the construction and surrounding terrain. Examples are provided that show effects of wind direction and speed and the geometry of buildings on surface wetting by rainfall and also deposition of corrosive chloride.

MICRO-ENVIRONMENT

The micro-environment becomes specific to a single built infrastructure, relating to the exposure environment sourced at specific exposed surface(s) comprising the infrastructure. When considering a specific building, pipeline, wharf or other infrastructure, the many built components will each be exposed to different and unique micro-environments. It is, therefore, important that built components are clearly defined and categorised into subgroups that are based on environmental exposure. Each component needs to be considered in the context of specific micro-environments and potential for ageing. For a new project that is at the design stage, the many components of the infrastructure to be built can be assigned particular exposures based on analysis of the micro-environments. This will be further discussed in the context of the design process within Chapter 6. This is illustrated in the following examples.

Microclimates within a bridge

The example shown in Figure 4.1 illustrates the different micro-environments that can be assigned to the superstructure of a bridge located at Gimsøystraumen in Norway (Haagenrud, 2003, 2004) and exposed to a

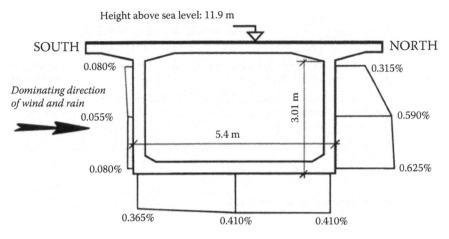

Figure 4.1 Cross-section of Gimsøystraumen Bridge in Norway, showing the influence of microclimates on the amount of surface chloride (% weight of concrete) deposited on the concrete surfaces. (From Figure 2.38, page 42, Gimsøystraumen Bridge, *Durability Design of Concrete Structures in Severe Environments*, Gjorv, O.E. (Ed.r), CRC/Taylor & Francis Group, 2014.)

severe marine environment. The cross-section of the bridge illustrates the chloride surface deposition that changes with geometry of exposed surfaces as well as the effect of the direction of wind-borne chloride. The microclimates are very different. The dominating direction of wind and rainfall produces a surface washing removal of chloride, whereas the leeward exposed face shows over ten times the surface chloride.

Figure 4.1 shows a cross section of a cantilever box girder on the bridge, composed of post-tensioned reinforced concrete. Under the influence of strong winds and breaking waves, seawater aerosol is transported, strongly influenced by the direction of the salt-laden prevailing wind. Surface chloride measurements were undertaken at nine different locations on the bridge girder, in a testing regime that included box girders located at different heights above the sea. The investigators concluded that corrosive chloride deposited on the surface is 6.5 times lower when the superstructure of the bridge rises from 8 to 28 metres above sea level. Interestingly, the highest surface chloride is eight times higher on the leeward vertical surface than on the opposite windward face of the box girder. Although the wind is directed at the windward vertical face, wind-driven rain washes some chloride from the surface. Following contact with the windward surface, wind-borne chloride is deflected, and the slower slipstream ensures higher deposits of chloride at the soffit, while a back-flowing slower eddy current of air at the leeward side produces considerably higher chloride deposition at the leeward face. The implications are higher risk of corrosion of the steel reinforcement on the leeward side. Consideration of micro-environments prior to construction can lead to improved geometry and configurations of built components, choice of building materials, construction methods and supplementary protection that can reduce the effects of ageing. More sophisticated methods of simulating dispersion modelling within multiple structures can be obtained by software including CALPUFF View™, which includes 3D complex terrain algorithms, calculated turbulence, shoreline boundaries and utilises 3D-computational modelling to visualise the entire surveyed area (Rajni et al., 2006). Salt transportation and deposition inland is less reported; however, it is an important meso/microclimatic effect in Brazil (Meira, 2006).

Microclimates within a wharf

Figure 4.2 shows contrasting microclimates, as illustrated by a wharf showing the many corrosive exposures:

1. The *Atmospheric Zone* is classified above peak wave height. Airborne corrosive chloride is transported over the exposed wharf surfaces with an intensity that lessens with increasing distance inland. Depending on the configuration and height of the wharf, and whether there exist any built fixtures that obstruct or deflect the prevailing winds, corrosive salt droplets and aerosol will be deposited. Exposed steel will be oxidised, whereas reinforced concrete has a time lag to corrosion due to diffusion of chloride through the concrete to the embedded steel. The airborne zone is less corrosive than the splash zone due to lower amounts and rates of chloride deposition as well as discontinuous wetting caused by the variable nature of rainfall and/or dewfall. Regardless, the function of the wharf influences exposure. A container wharf will have areas on the deck slab that are intermittently sheltered from chloride due to stacking and restacking of containers, whereas a passenger terminal has mostly unprotected zones. Surfaces exposed to the leeward face of prevailing winds (e.g. vertical wall of a storage building) can collect higher deposits of chloride than the windward wall due to deflection of winds around the building and the resultant slowing of the wind on the leeward wall.
2. The *Splash Zone* exposes the wharf to high surface chloride under wetting and drying conditions. Corrosion of exposed steelwork generally proceeds at steady to fast rates due to the availability of plentiful oxygen as well as wet/humid conditions. Conditions will vary with the intensity of prevailing winds and shelter provided by a protective harbour, and the proximity and configuration of adjacent port infrastructure (berths, cranes, storage buildings and other fixtures). Even when 'dry', the hygroscopic nature of chloride salts ensures a damp surface that is conducive to corrosion. Figure 4.2 shows more intense splash at the rear of the wharf due to the configuration of the

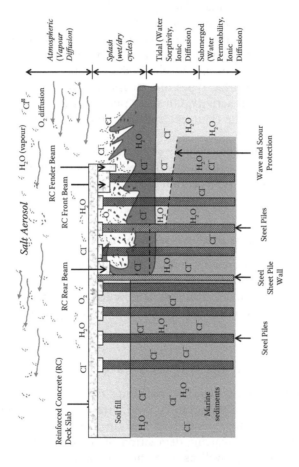

Figure 4.2 Cross-section of a wharf showing the different exposure zones. (Courtesy of F. Collins.)

structural members; highest splash occurs at the rear where waves impact on the sheet pile wall and rebound upward, resulting in exposure of the undersides of the wharf. This contrasts with the centre of the wharf. Similarly, a high zone of splash occurs at the front of the wharf due to waves impacting the fender beam as well as swash arising during berthing of ships.

3. The *Tidal Zone* fluctuates within the day, ranging in height from immersion in seawater to air-exposed, while varying in height and frequency with calendar date, geographical location, sea currents and winds. Two areas of concern are the high tide level where there exists sufficient oxygen to support corrosion, while the second concern arises at the low water height of the tide where accelerated low water corrosion can occur. Accelerated low water corrosion can occur more rapidly that other types of corrosion, evidenced by pitting arising from microbial activity. On sheet piles walls, concealed corrosion can occur due to the chemical and microbiological characteristics of the soil behind.

4. There is a lack of oxygen within seawater at depth, and the *Submerged Zone* is protected at the steel surface by marine growth that forms a film at the exposed surface. In most cases this limits the rate of corrosion to minimal.

Underground within marine sediments

Due to being saturated by seawater, the first conclusion may be a low corrosion risk due to the low oxygen content of the sediments. However, in this anaerobic environment, bacteria (e.g. sulphate-reducing) can induce microbiological corrosion to the steel piles and sheet pile wall if there are sufficient nutrients.

Urban influences on the microclimate

Urban 'canyons' between high-rise buildings can influence air flow and wind speed; conversely a topographic configuration of many buildings of different heights can cause wind drag and turbulence that cause variable exposure conditions, influencing deposition of corrosives (airborne acidic gases and/or chloride) as well as local heating and cooling and wetting/drying.

Construction materials generally have lower specific heat than vegetation and soils. Hence, during the day, less energy is needed to raise the temperature of built infrastructure when compared with adjacent rural areas, causing the so-called 'Heat Island' effect where the urban area has significantly higher temperature (2°C–12°C) than the air temperature of nearby rural areas (Arnfield, 2003). Higher temperatures accelerate the deterioration processes of most materials.

Considerations that can significantly affect local surface exposures include:

1. Exposure to solar radiation (sun path and influence of orientation, exposed/sheltered/filtered, access to openings, severity and for how long).
2. Temperature conditions (the component located in an urban environment where there exists urban heat island effects).
3. Wind exposure to chloride and/or acidic pollutants and whether the particular component has an orientation and geometry that is conducive to high/lower deposition of corrosives. Effect of adjacent terrain (natural and man-made) needs consideration.
4. Will the component be surface-wet for extended periods due to high exposure to rainfall and minimal drying?

In conclusion, the macro-, meso- and micro-environments are multiple types of exposures that have a key bearing on the durability of the built infrastructure. This chapter has characterised the different environmental exposures, or more correctly, the *durability loading* that is imposed on infrastructure.

REFERENCES

Arnfield, A. J. (2003). Two decades of urban climate research: A review of turbulence, exchanges of energy and water, and the urban heat island, *International Journal of Climatology*, 23(1): 1–26.

Beavers, J. (2001). State-of-the-art survey on corrosion of steel piling in soils, *NACE International Task Group 018*. Houston, TX: NACE International.

BGS. (2015). Corrosivity map for the UK. British Geological Survey. http://data.gov. uk/dataset/corrosivity-map-for-the-uk.

Blumthaler, M. and Ambach, W. (1988). Solar UVB–Albedo of various surfaces, *Photochemistry and Photobiology*, 48(1): 85–88.

Boyd, G. M. (1970). *Brittle Fracture in Steel Structures*. London, UK: Butterworth and Co., 119p.

BRE. (2005). *Concrete in Aggressive Ground*. Watford, UK: Building Research Establishment, 68p.

Carlsson, T. (2003). Report D4.3 GIS-based national exposure modules and national reports on quantitative environmental degradation loads for chosen objects and locations, LIFECON Life Cycle Management of Concrete Infrastructures for Improved Sustainability, G1RD-CT-2000-00378, 37p.

Chin, J. W. (2007). Ch5–Durability of composites exposed to ultraviolet radiation. In: V. M. Karbhari (Ed.), *Durability of Composites for Civil Structural Applications*. Sawston, UK: Woodhead Publishing, pp. 80–97.

Cole, I. S., King G. A., Trinidad G. S., Chan W. Y., and Paterson, D. A. (1999). An Australia-wide map of corrosivity: A GIS approach. Australia-wide map of corrosivity, *CIB W078 Workshop on Information Technology in Construction*. Rotterdam, the Netherlands: In House Publishing, 11p.

Cole, I. S., Paterson, D. A., and Ganther, W. D. (2003). Holistic model for atmospheric corrosion Part 1 – Theoretical framework for production, transportation and deposition of marine salts, *Corrosion Engineering, Science and Technology*, 38(2): 129–134.

Dolske, D. A. (1995). Deposition of atmospheric pollutants to monuments, statues, and buildings, *Science of the Total Environment*, 167(1–3): 15–31.

Doran, S. R., Robery, P., Ong, H., and Robinson, S. A. (1987). Corrosion protection to buried structures, *Construction and Building Materials*, 1(2): 88–97.

E-PRTR (2015). European pollutant release and transfer register. http://prtr. ec.europa.eu/.

Feliu, S., Morcillo, M., and Chico, B. (1999). Effect of distance from sea on atmospheric corrosion rate, *Corrosion*, 55(9): 883–891.

Gjorv, O. E. (Ed.) (2014). *Durability Design of Concrete Structures in Severe Environments*. Boca Raton, FL: CRC Press/Taylor & Francis Group.

Guo, Y., Shi, Z., Li, H. Y., and Triantafilis, J. (2013). Application of digital soil mapping methods for identifying salinity management classes based on a study on coastal central China, *Soil Use and Management*, 29(3): 445–456.

Haagenrud, S. (2003). *LIFECON Life Cycle Management of Concrete Infrastructures for Improved Sustainability Project*, Helsinki, Finland, 65p.

Haagenrud, S. E. (2004). Instructions for quantitative classification of environmental degradation loads onto Structures, *LIFECON Life Cycle Management of Concrete Infrastructures for Improved Sustainability Project Report* 378: 63p.

Haque, A. M., Al-Khaiat, H., and John, B. (2007). Climatic zones—A prelude to designing durable concrete structures in the Arabian Gulf, *Building and Environment*, 42(6): 2410–2416.

Huang, J., Nhan, T., Wong, V. N. L., Johnston, S. G., Lark, R. M., and Triantafilis, J. (2014). Digital soil mapping of a coastal acid sulfate soil landscape, *Soil Research*, 52(4): 327–339.

IGBP-DIS. (1998). SoilData (V.0) A program for creating global soil-property databases. IGBP Global Soils Data Task, France. http://atlas.sage.wisc.edu/.

IODE. (2015). International oceanographic data and information exchange. http:// www.iode.org/.

IUGS. (2015). International union of geological sciences (IUGS). http://iugs.org/.

Javaherdashti, R. (2011). Impact of sulphate-reducing bacteria on the performance of engineering materials, *Applied Microbiology and Biotechnology*, 91(6): 1507–1517.

King, G. A. and Moresby, J. F. (1982). *A Detailed Corrosivity Map of Melbourne*. Highett, Australia: CSIRO, 13p.

Lewis, E. R. and Schwartz, S. E. (2013). *Fundamentals. Sea Salt Aerosol Production: Mechanisms, Methods, Measurements and Models—A Critical Review*. Washington, DC: American Geophysical Union, pp. 9–99.

Marsh, W. M. and Kaufman M. M. (2012). Global climate types and descriptions, *Physical Geography*. Cambridge, UK: Cambridge University Press, 634p.

Meira, G. R., Andrade, M. C., Padaratz, I. J., Alonso, M. C., and Borba, J. C. (2006). Measurements and modelling of marine salt transportation and deposition in a tropical region in Brazil, *Atmospheric Environment*, 40(29): 5596–5607.

Myer, E. (2013). *Plastics Failure Guide – Cause and Prevention*, 2nd ed. Munich, Germany: Hanser, 833p.

Naylor, S. D., Chapman, G. A., Atkinson, G., Murphy, C. L., Tulau, M. J., Flewin, T. C., Milford, H. B., and Morand, D. T. (1998). *Guidelines for the Use Acid Sulfate Soils Risk Maps*, 2nd ed. Report, Department of Land and Water Conservation, Sydney, Australia, 25p. http://www.environment.nsw.gov.au/resources/acidsulfatesoil/assmapsguide.pdf.

NODC. (2015). US national oceanographic data center. http://www.nodc.noaa.gov/.

NPI. (2015). National pollutant inventory. http://www.npi.gov.au/.

Okabe, S., Odagiri, M., Ito, T., and Satoh, H. (2018). Succession of sulfur-oxidizing bacteria in the microbial community on corroding concrete in sewer systems, *Applied and Environmental Microbiology*, 73(3–4): 971–980.

Peel, M. C., Finlayson, B. L., and McMahon T. A. (2007). Updated world map of the Köppen-Geiger climate classification, *Hydrology and Earth System Sciences Discussions*, 11(5): 1633–1644.

Rahman, M. M. and Bassuoni, M. T. (2014). Thaumasite sulfate attack on concrete: Mechanisms, influential factors, and migration, *Construction and Building Materials*, 73: 652–662.

Rajni, O., Kumar, A., and Masuraha, A. (2006). Application of the USEPA's CALPUFF model to an urban area, *Environmental Progress and Sustainable Energy*, 25(1): 12–17.

Rasheeduzzafar, R., Fahd, H. D., and Al-Gahtani, A. S. (1984). Deterioration of concrete structures in the environment of the Middle East, *Applied Clinical Informatics Journal*, 81(1): 13–20.

Rengasamy, P. (2006). World salinization with emphasis on Australia, *Journal of Experimental Botany*, 57(5): 1017–1023.

Robery, P. C. (1988). Protection of structural concrete from aggressive soils, *Chemistry and Industry (London)*, 13: 421–426.

Sabbioni, C. (2003). Mechanisms of air pollution damage to stone, *The Effects of Air Pollution on the Built Environment*, 2: 63–106.

Siad, H., Lachemi, M., Bernard, S. K., Sahmaran, M., and Hossain, A. (2015). Assessment of the long-term performance of SCC incorporating different mineral admixtures in a magnesium sulphate environment, *Construction and Building Materials*, 80: 141–154.

Silver, W. L., Lugo, A. E., and Keller, M. (1999). Soil oxygen availability and bio-geochemistry along rainfall and topographic gradients in upland wet tropical forest soils, *Biogeochemistry*, 44(3): 301–328.

Slamova, K., Glaser, R., Schill, C., Wiesmeier, S., and Köhl M. (2012). Mapping atmospheric corrosion in coastal regions: Methods and results, *Journal of Photonics for Energy*, 2(1): 11p.

Toth, G., Adhikari, K., Várallyay, G., Toth, T., Bodis, K., and Stolbovoy, V. (2008). Updated map of salt-affected soils in the European Union. In: Toth, G., Montanarella, L. and Rusco, E. (Eds.), *Threats to Soil Quality in Europe*. Ispra, Italy: JRC European Commission, 13p.

USGS (2015). US geological survey (USGS). http://ngmdb.usgs.gov/.

WMO (2015). World meteorological organization. http://www.wmo.int/.

Chapter 5

Predictive modelling of ageing

'Prophesy is a good line of business, but it is full of risks' — Mark Twain in *Following the Equator*

INTRODUCTION

Chapters 3 and 4 reviewed the mechanisms and the effect of the exposure environment on ageing of infrastructure. An important aspect of asset management is the ongoing gathering of data and the proactive maintenance planning toward improving the durability of infrastructure. What is to be done with this voluminous data, including construction records, ongoing regular condition surveys and inspections, servicing and maintenance records? At your peril, 'do nothing' seems an easy option (although an essential option due to either lack of maintenance funding or that the infrastructure is not of critical risk), and it could lead to premature and unexpected maintenance and rehabilitation and possibly early replacement of the built infrastructure. Nevertheless, asset managers may be constrained by funding restrictions, or alternatively the particular asset may not be critical and therefore is deemed a low priority for future maintenance.

The major cost during the early part of the asset life cycle is in the construction, although the capital cost depreciates each year while maintenance and remediation costs accrue. As the assets age, the maintenance cost accrues until it reaches and then exceeds the construction cost (e.g. if annual maintenance costs are 2 percent of the initial capital cost, then the maintenance cost equals the capital cost after fifty years of Service Life). These maintenance costs may be higher or lower depending on the accessibility of the infrastructure as well as the priority of the asset. For example, buildings or pipes that are difficult to access can be expensive to operate and maintain compared with a more accessible low-level bridge. The operating costs by comparison *should* be relatively small however the often-unconsidered costs relate to the ongoing deterioration and remediation of the asset due to exposure to corrosives, as well as usage 'wear and tear'. If allowed to deteriorate without due diligence to upkeep and maintenance, the asset may lose its ability to function cost-effectively, or even to function at all due to premature failure. Conversely, vigilant construction and maintenance could extend the Service Life of the asset that is greater than the Design Life, a favourable and cost-effective outcome. According to Wong and de Almeida (2014) in World Economic Forum Report 180314: 'As a result of the maintenance backlog and the lack of resilience measures, existing assets deteriorate much faster than necessary, shortening their useful life.... For example, lack of an effective Operation and Maintenance System (O&M) leads to congestion and unproductive use capacity, poor quality for users...a further example involves a large-scale power outage in India left approximately 700 million people without electricity in July 2012, cost inefficiencies (e.g. crane productivity at many ports is only half of the over 40 moves per hour that world-class ports can achieve), and environmental and social externalities (e.g. more than 25 million metric tons of CO_2 are produced in US traffic jams each year)'. In contrast, Wong and de Almeida (2014) also quoted the Panama Canal, an 80 km connection between the Atlantic and Pacific Oceans, which manages to maximise asset utilisation, maximise quality for users, reliability of service and maintenance planning scheduled to avoid customer waiting times, thereby optimising efficiency.

WHY UTILISE PREDICTIVE MODELS?

Predictive models utilise variable asset inspection and test data and statistical probability toward forecasting the future condition and risk of the asset. Forecasting the future condition of the asset takes into account the uncertainties associated with the accounting for asset condition that is subsequently accounted for by a risk analysis. This enables putting into place the best intervention options toward controlling the ageing of infrastructure and, therefore, lowers the risk of loss of function and premature closure of the facility. Predictive modelling of performance supports the management questions of when and/or how the deterioration occurred as well as to proactively circumvent potential problems.

TYPES OF INFRASTRUCTURE PREDICTIVE AGEING MODELS

Predictive modelling of ageing of infrastructure is a powerful tool toward more proactive maintenance and optimisation of life-cycle costs. The types of predictive modelling include:

Sit and wait

Essentially no inspection, testing and predictive modelling is undertaken, followed by ageing of a component (s) of the infrastructure. This result could be acceptable on a non-critical component of the infrastructure or infrastructure that is designed specifically for a short life span; however, on critical components this can result in unanticipated rehabilitation or premature replacement as well as follow-up costs involving reduced functionality, safety and efficiency.

Empirical

Empirical analysis is based on verifiable information observed by the outcomes of site and/or laboratory experiments and/or experience and/or observations. It does not solely rely on the mechanisms of behaviour and scientific theory alone, often without due regard for system and theory. Empirical analysis can provide useful and significant predictions when coupled with multiple regression analysis, for example, from a sampling 'health check' survey that provides information on deterioration trends within a group of port structures. Prudence is needed when choosing the definition of the problem and of the parameters to be assessed. The sampling can be extended, for example, to identify the deterioration effects of geometrical layout, geographical location, sea conditions and amount of shelter; however, the trends and outcomes are only as good as the quality and volume of data taken and the wisdom of the choice of sampling locations.

Deterministic

The deterministic methodology is based on a sequence of mathematical steps that are strictly followed in the analysis. Deterministic predictive models return a given value of condition following ageing based on a set of mathematical equations determined through known relationships between parameters (e.g. properties of construction materials) contributing to ageing. There is no accounting for random variation within parameters. For example, the rate of corrosion of steel may be described by the power law:

$$C = kt^n$$

where:
    C represents corrosion loss (g/m^2)
    k is the corrosion loss within the first year
    t is the exposure time in years
    n is a coefficient that characterises the protective properties of corrosion products

The values of k and n are derived by calibration of the equation to known test data. Although the power law allows quick estimation of the rate of deterioration by corrosion, it is vulnerable to the need for sound test data based on a particular metal, initial condition and exposure.

Deterministic tools do not consider the uncertainty inherent in durability assessments, nor do they explicitly address the potential risks. Nevertheless, allowance is made during the calculations for the uncertainties of the basic variables by means of characteristic values and partial factors.

Mechanistic

This method assumes the behaviour of complex systems can be assessed by analysis of the interrelated individual components comprising that system. For example, the deterioration of timber may be analysed by the components leading to deterioration (e.g. species and characteristics of the timber, porosity, moisture content, exposure type and their interrelationships toward deterioration). In a world where deterioration of infrastructure can occur randomly or in unexpected locations, the mechanistic cause-and-effect definition can be sometimes rigid and unlikely to plan and avoid future failure and maintenance. Nevertheless, in order to predict bridge deterioration in advance, Colorado State University (Nickless and Atadero, 2017) implemented mechanistic models on bridge structures that analysed the physical processes causing deterioration while addressing limitations associated with bridge inspection data and statistical methods. Due to its prevalence throughout the state of Colorado and frequent need for repair, corrosion-induced cracking of reinforced concrete (RC) decks was selected as the mode of deterioration for further study. Probabilistic inputs were applied to simulate inherent randomness associated with deterioration. Model results showed that mechanistic models may be able to address the limitations of deterministic models and provide a more accurate and precise prediction of bridge degradation in advance. Statistical models such as a Markov Chain or Weibull distributions can be utilised, relying on past data from, say, biannual visual inspections to predict future deterioration. Alternatively, mechanistic models attempt to predict the condition of a bridge by analytically describing the physical mechanisms causing deterioration. They use environmental and other physical data such as concrete mix parameters as inputs to predict how a bridge element will degrade over time. The complicated nature of multiple deterioration mechanisms presents a challenge for creating accurate mechanistic models. In the Colorado study, analytical models that represent the individual stages of deterioration of RC bridge decks were combined and modified to reflect current design practices (e.g. protective epoxy-coated reinforcing steel, waterproofing membranes and asphalt wearing surfaces). Interactive effects between bridge decks and joints were also reported to demonstrate the ability of mechanistic models to predict deterioration of multiple elements simultaneously. Then the effects of maintenance actions on model outputs were examined. Rather than a replacement for current statistical models in use by departments of transportation, mechanistic models may be used as a supplement to fill in gaps where condition information is missing.

Probabilistic

The majority of factors that are used for an empirical or deterministic analysis are generally static in nature, whereas the probabilistic approach allows a degree of uncertainty of each of the factors when undergoing mathematic predictive analysis. Probabilistic state-based/time-based models are used to predict the global or macro-response of built components for network level analysis. In the probabilistic analysis, input values are randomly generated; based on the users' statistical distributions (e.g. values measured during condition surveys of built infrastructure), the probability of failure can be calculated. The calculations are iterative, with each calculation based on a new random value chosen from the statistically defined input factors (e.g. the Monte Carlo method randomly and iteratively selects values from the distributions based on each of the input factors). The results from the analysis are utilised to predict two possible types of failures: (a) failure occurs when a cumulative amount of damage or a deterioration process exceeds a permissible limit; and (b) failure occurs when an engineering parameter exceeds the corresponding capacity. A factor of safety is preset to confirm the estimated probability of durability failure is less than the factor of safety, that is, an effective means of controlling potential risk.

The Bayesian method is often applied within the probabilistic analysis and is a means of updating the probability for each of the factors (prior distribution) as more evidence or information becomes available during the probabilistic analysis. For example, the Bayesian method can utilise and update visual damage condition records of, say, corroding pipelines, to provide improved forecast on the likelihood and geographical location of future pipeline failures. Other probabilistic approaches currently employed for durability analysis include Fuzzy Logic (based on 'degrees of truth' rather than the Boolean logic 'true or false' [1 or 0]) and Markov Chains that describe durability based on a series of prior related and probabilistic events.

Reliability of ageing infrastructure

Reliability-based mechanistic models are used to predict the detailed or micro-response of built components for project level analysis. Reliability-based mechanistic models are developed using quantitative performance indicators (physical parameters) that are determined through, for example, detailed condition surveys, analytical modelling and empirical investigations to identify the extent and severity of specific deterioration. For a pre-designated risk of failure, the deteriorated components can be assessed. ISO2394 (2015) provides a rigorous coverage on the general principles of risk and reliability for structures. The standard provides not only the design and construction phases of built infrastructure but also the operation, maintenance, rehabilitation and decommissioning. The International Standard provides for approaches at three related levels, namely the following:

1. Risk informed, that is, the effect of uncertainty on the built item. Prior to the 1960s, there was ignorance about the corrosive effects when built infrastructure was exposed to salts (e.g. chloride-based deicing salts applied to bridge decks undertaken to melt snow). In recent years, the risk has been identified and dealt with (although the subject of ongoing debate); however, the deterioration legacy from that infrastructure remains and has led to significant rehabilitation. Risk analysis aims to balance lowest/acceptable risk with existing resources to achieve the most durable built component for the intended level of infrastructure service. Put simply, risk of deterioration can be expressed as:

$$\text{Risk} = \text{Probability of Unacceptable Deterioration } (P_d)$$

$$\times \text{Consequences of Failure } (C_f)$$

Consequences of failure have been discussed in Chapters 1 and 3.

2. Reliability based, that is, the ability of the built component to fulfil its designed performance requirements during the Service Life (usually expressed as a probability). In general terms, the resistance to deterioration of a built component (R) is greater than the applied corrosives to the built components (S), or $R \geq S$.

   The reliability method is a means of providing discrete probability distributions, with degrees of variation, to each of the variables that comprise R and S. The net probability can then be estimated, whether $R > S$ or $R < S$, allowing for a margin of safety. Each variable has a probability of behaviour, which can be defined, for example, by binomial, Weibull, Poisson or other discrete distributions. However, when there are multiple variables to be considered, the problem is less simple and a multivariate approach is needed. Assuming, say, all the variables have a normal distribution, the bivariate normal distribution assumes the variables relate as a joint probability distribution. Where there are more variables, the bivariate distribution can be extended to the multivariate normal distribution.

   A life cycle profile deals with events and conditions associated from the time of construction completion to the end of life of the infrastructure. The ageing reliability of the infrastructure depends on exposure to corrosives, magnitudes of stresses and geometrically where these actions are situated. Examples of corrosive exposures could include time-of-wetness for a steel surface, proximity to airborne chloride and exposure to acidic gases. The initial step is to describe the expected deterioration events, both environmental and physical, throughout the life cycle, followed by quantifying the loading conditions from available data from life-size or laboratory measurements in a statistical manner to identify the size and variability of the loadings.

3. Semi-probabilistic (or partial factor), that is, for structures that the consequences of failure and damage are well understood and the failure modes can be categorised and modelled in a standardised manner; allowances for uncertainties can be assigned to basic variables by means of representative values and partial factors.

The level of reliability is related to the possible consequences of failure and necessary procedures to reduce the risk of failure and damage. It is expected that the built component will retain functionality through the Service Life by withstanding deterioration to within a target level (ideally agreed at the time of Design Life).

Multi-scale

Construction materials, for example concrete, are not homogeneous materials but have a heterogeneous internal structure comprising of coarse and fine aggregates that are bound by a cementitious binder. Similarly, timber consists of fibres, cellulose, lignins, and pores that constitute the heartwood and sapwood of a tree to be utilised for timber construction. The behaviour of built concrete components of infrastructure is complex, and the behaviour can be characterised from atomic-scale to macro-scale. One central problem in macro-scale simulations is describing the overall (homogenised) response of concrete and the determination of the corresponding effective material parameters (Koichi et al., 2009). Figure 5.1 shows the multiple range of sizes within concrete, including the void network ranging from atomic-sized gel pores (and impregnable) to macro-scale capillary pores and sizes that are visible to the eye. When considering transport of corrosives through concrete, the gel pores have a minor role to play, whereas capillary pores (and larger air voids) have a more significant role. In most macro-scale predictive models, concrete is considered to be a homogeneous material; however, the range of pores and microcrack influences are not considered; rather a macrosize is generally adopted in durability calculations. Figure 5.1 shows the multiple ranges of sizes within concrete and how water and entrapped air contribute to the size range.

Unger and Eckardt (2011) present a meso-scale model of concrete, which considers particles, matrix material and the aggregate-paste interfacial transition zone (ITZ) as separate constituents, the purpose being to interlink heterogeneous scale models with the macrobehaviour of the concrete. Algorithms are proposed to generate realistic particle configurations efficiently. The nonlinear behaviour of concrete can be attributed to the initiation, propagation, accumulation and coalescence of microcracks within microstructure and macrostructure of the material. Thus, failure of concrete structures is a multiscale phenomenon – the material behaviour of concrete on the macro-scale, which

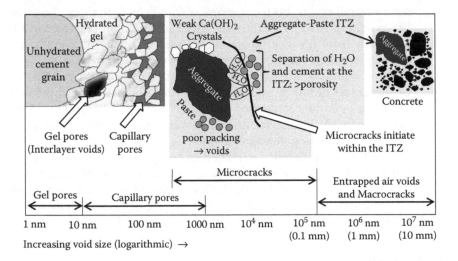

Figure 5.1 Multi-scale contrast within concrete, showing atomic-sized gel pores through to considerably larger macrocracks and air voids. (Courtesy of F. Collins.)

corresponds to the structural level, is clearly influenced by the geometry, the spatial distribution and the properties of the individual material constituents and their mutual interaction (Michel et al., 2016). A localisation of these microcracks, which is also triggered by the heterogeneous internal material structure, leads to the formation of macroscopic cracks and is accompanied by a softening of the material. In meso-scale simulations, the numerical model explicitly represents the individual components of the heterogeneous internal material structure of concrete, for example, the shape and the spatial distribution of the aggregates. As a result, specific material models can be assigned to each component of the meso-scale model. These models also represent phenomena, such as size effects on the nominal tensile strength, the stochastic scatter of the results in simulations with specimens with identical shape but different aggregate configurations or the localisation of damage due to the heterogeneity of the material.

This is further explored by Maekawa et al. (2003) who couples the microstructure, cementitious hydration and moisture as an integrated approach. Chemical, mechanical and related stress models that explain concrete behaviour in 3D, with emphasis that the micro-scale behaviour is mutually interrelated with thermodynamic events (dynamic coupling of pore-structure development to both the moisture transport and hydration models, the development of strength along with moisture content and temperature can be traced with the increase in the degree of hydration for any arbitrary initial and boundary conditions). The methodology accounts for dynamic changes to the cement particle size distribution. Many of the smaller particles dissolve completely, leaving room for the larger particles to grow as the hydration proceeds; although, Maekawa has made the simplification of the cement particles being of circular geometry. As the hydration proceeds, new products are formed and deposited in the large water-filled spaces known as capillary pores. Moreover, the gel products also contain interstitial spaces, called gel pores, which are at least one order of magnitude smaller than the large voids of capillary pores. The size of pores present in the cement paste is distributed over several orders of magnitude, and the distribution of moisture in such a porous system is different for each different range of pore sizes considered. Thus, the pore structure formed in the hydrated mass is highly complex due to its complicated shape, interconnectivity and spatial distribution.

Multiple numerical models describing different length scales are combined. The development of the pore structure at early ages utilises a model based on the average degree of hydration and used as a basis for moisture transport computation. The contrasting geometrical scales are represented as reference control volumes (RCV) with each assigned mathematical equation that describes the material behaviour. The events are interlinked to describe overall macrodeterioration as a function of simultaneous geometrical, chemical, mechanical and diffusion behaviour. An important aspect of the meso-scale simulation of concrete is the accurate representation of the heterogeneous material microstructure characterised by particle shapes, their size distribution according to a prescribed grading curve and the spatial position and orientation of the particles within the specimen. The first possibility is based on image processing techniques. Based on X-ray computer tomography or by sequential sectioning and two-dimensional image processing, a three-dimensional voxel representation of the microstructure is obtained, which can be used in a voxel-based finite element representation.

Damage simulation and visualisation in 3D

At present, infrastructure ageing prediction techniques, for example, for RC are based largely on simulation models that focus on predicting the initiation of 1D or 2D corrosion of steel within reinforced concrete. While some propagation phases (that is, from corrosion initiation onward) can lead as far as structural deterioration, the modelling is still at a developmental stage. Existing models simplify the RC geometry via 1D or 2D thick-walled cylinder models, which can only predict damage for a single cross section of a structure and lack the capability to predict damage patterns that vary along the third dimension (or length) of structural members in real life. Furthermore:

(1) corrosion rate is often assumed to be constant over time and (2) corrosion products are uniformly distributed around the reinforcement bar circumference and along the third dimension (or length) of the structural member. These assumptions greatly help to simplify the modelling process; however, these assumptions do not accurately represent the corrosion process in field RC marine structures.

Chen and Mahadevon (2008) modeled chloride ingress followed by corrosion-induced cracking due to expansive internal reinforcement corrosion.

Progress was made with a 3D finite element model that is based on a smeared cracking approach, where many small regions of high strain are representative of the coalescence of microcracks into macrocracks that progressively propagate. The authors utilized a 3D finite element to simulate the progressive expansive impacts due to steel corrosion and were able to simulate crack propagation for the occurance of both spalling and delamination within a slab structural component. The 3D application assumes constant conditions within the 3rd dimension (length) while the fixed geometry is not easily adapted for different graphical geometries.

Ozbolt et al. (2011) undertook a 3D approach, whereby the formation of a crack following tensile loading, and perpendicular to the longitudinal steel reinforcement, resulted in the depassivation of steel. The crack serves as the conduit for corrosive media to reach the embedded steel reinforcement. The outcome from the work was that existing tensile cracks do not greatly influence the corrosion rate. Ozbolt et al. (2013) progressed the 3D model to predict the crack damage patterns arising following depassivation of steel reinforcement. Anodic and cathodic positioning was varied along the beam length in the numerical analysis. Reasonable comparison between predicted (2D) and laboratory-exposed samples under accelerated conditions was achieved. Nevertheless, the effect of the resistivity between anode and cathode as well as changing exposure conditions varying along the length and width of the beam needs consideration as well as the geometry of the diameter of steel bars to the cover to reinforcement. Ozbolt et al. (2016) progressed the 3D model to undertake analysis of reinforcement corrosion on beams with closely spaced steel bars, with and without vertical steel stirrups. The model allows for the distribution of anodic to cathodic regions around the circumference of steel bars and assumes the concrete within the cathodic regions are fully saturated with oxygen. A smeared crack approach was undertaken. Good agreement was achieved by comparison with experimental data. However, only 130 mm segments of the beam length were numerically analysed and therefore the possibility of anodic and cathodic regions along the beam were not considered.

Siamphukdee (2015) undertook 3D finite element analysis that dealt with time-variant corrosion rate, simulated non-uniform distribution of rust products around the reinforcement bar, and includes an environmental factor that allows for variation of corrosivity along the longitudinal dimension of the RC component being analysed. A smeared cracking approach was adopted to analyse a wide range of structural geometries, material properties, and exposure environments. The outputs from the numerical model were compared and contrasted with experimental test data as well as the performance of in-service reinforced concrete port structures. Good correlation was achieved between the 3D numerical model and the laboratory and site-based sample behaviour. Most predictive models assume a constant corrosion rate throughout the service life of a RC structure whereas Siamphukdee (2015) allows for variation of corrosion rate (leading to corrosion-induced pressure) according to the response of the exposed steel to varying environmental exposures along the length of the RC component. Two cross-sectional geometries were developed: (i) a corner geometry incorporating an embedded single bar with bi-directional chloride loading and unrestrained by the surrounding concrete within the exposed faces., and (ii) an embedded single non-corner bar that is subject to singular direction chloride loading and concrete unrestrained only at the exposed orientation. These elements are versatile and can be configured to achieve a range of cross-sections and reinforcement layouts. Exposure was varied in the longitudinal direction and therefore the model simulated the effects along the length of a RC component of a structure. Progressive buildup of corrosion products on embedded steel bars was evaluated around the bar circumference as well as length by 3D finite element analysis, followed by calculation of corrosion-induced stresses leading to predicted damages (e.g. cracking). Cracking was predicted at a particular element within the cover concrete when the predicted tensile strain exceeds the concrete tensile strain capacity or where the overall displacement is sufficiently large for cracks to occur. An example of the predicted damage to a single embedded bar (unrestrained and exposed to chloride in three directions) in Figure 5.2.

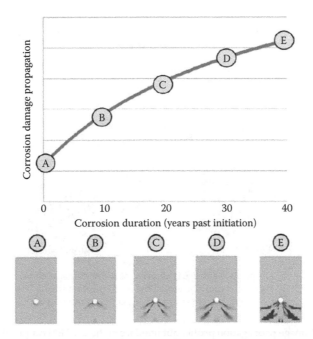

Figure 5.2 Major types of damage patterns identified by FEM. (From Siamphukdee, K., Development of 3D predictive ageing simulations of reinforced concrete port structures, PhD Dissertation, Department of Civil Engineering, Monash University, Clayton, Australia, 2015.)

The calculation of rust productions in the subsequent stages of corrosion is more complicated due to the difficulty to predict the type of corrosion products formed during the following corrosion process; the types of corrosion product formed are highly sensitive toward the environmental and electrochemical conditions within the reinforced concrete system. An example of a corrosion damaged beam is shown in Figure 5.3. The beams in Figure 5.3 show very different signs of corrosion damage and corresponding to differences in chloride content along the beam.

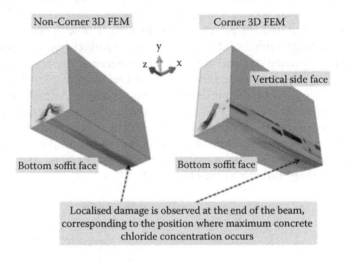

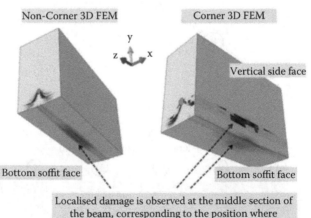

Figure 5.3 Damage propagation profile obtained from the 3D FEM output showing localised damage at the middle of the beam. (From Siamphukdee, K., Development of 3D predictive ageing simulations of reinforced concrete port structures, PhD Dissertation, Department of Civil Engineering, Monash University, Clayton, Australia, 2015.)

CONCLUSION

There are many available methods for determining time to damage or post-damage behaviour of structure elements, whether exposed to corrosive or physical or chemical exposure. The methods range from deterministic to probabilistic, mechanistic, multi-scale or single-scale – there is not a single methodology that can succinctly account for the variability of the exposure environment, composition of the engineering materials and the variations in geometry. This chapter has provided a range of alternatives, with analysis of the advantages and disadvantages of each.

REFERENCES

ABAQUS. (2011). *ABAQUS 6.11 Analysis User's Manual Dassault Systèmes*, Providence, RI. http://130.149.89.49:2080/v6.11/index.html.

Chen, D. and Mahadevan, S. (2008). Chloride-induced reinforcement corrosion and concrete cracking simulation, *Cement & Concrete Composites*, 30: 227–238.

ISO. (2015). *International Standard ISO 2394 General principles on Reliability for Structures*, 4th ed. Geneva, Switzerland: International Organisation for Standardisation, 111p.

Koichi, M., Ishida, T., and Kishi, T. (2009). *Multi-Scale Modeling of Structural Concrete*. Boca Raton, FL: CRC Press, 658p.

Maekawa, K., Ishida, T., and Kishi, T. (2003). Multi-scale modeling of concrete performance—Integrated material and structural performance, *Journal of Advanced Concrete Technology*, 1(2): 91–126.

Michel, A., Stand, H., Lepach, M., and Greiker, M. R. (2016). Multi-physics and multi-scale deterioration modelling of reinforced concrete, *Key Engineering Materials*, 665: 13–16.

Nickless, K. and Atadero, R. (2017). Research Report CDOT-2017-05 Investigation of mechanistic deterioration modeling for bridge design and management, Colorado Department of Transportation, in corporation with US Department of Transportation, Federal Highway Administration, 135pp. https://www.codot.gov/programs/research/pdfs/2017-research-reports/2017-05.

Ozbolt, J., Balabanic, G., and Kuster, M. (2011). 3D numerical modelling of steel corrosion in concrete structures, *Corrosion Science*, 53: 4166–4177.

Ozbolt, J., Orsanic, F., and Balabanic, G. (2016). Modeling corrosion-induced damage of reinforced concrete elements with multiple-arranged reinforcement bars, *Materials and Corrosion*, 76(5): 542–552.

Ozbolt, J., Orsanic, F., Kuster, M., and Balabanic, G. (2013). Modeling damage of concrete caused by corrosion of steel reinforcement, In: van Mier et al. (Ed.), *Proceedings VIII International Conference on Fracture Mechanics of Concrete and Concrete Structures*, FraMCoS-8, Toledo, Spain, March 11–14, pp. 1–12.

Siamphukdee, K. (2013). Sensitivity analysis of corrosion rate prediction models utilized for reinforced concrete affected by chloride, *Journal of Materials Engineering and Performance*, 22(6): 1530–1540.

Siamphukdee, K. (2015). Development of 3-D predictive ageing simulations of reinforced concrete port structures, PhD Dissertation, Department of Civil Engineering, Monash University, Clayton, Australia.

Unger, F. U. and Eckhardt, S. (2011). Multi-scale modeling of concrete, *Archives of Computational Methods in Engineering*, 18: 341–393.

Wong, A. and de Almeida, P. R. (2014). Strategic infrastructure steps to operate and maintain infrastructure efficiently and effectively, World Economic Forum, REF 180314, 88p. http://www3.weforum.org/docs/WEF_IU_StrategicInfrastructureSteps_Report_2014.pdf.

Chapter 6

Whole-of-life engineering for ageing infrastructure

'If you think good design is expensive, you should look at the real cost of bad design' – Sir Michael Bichard, The Design Council Chair (UK)/Ralf Speth, CEO of Jaguar Land Rover

INTRODUCTION

For infrastructure to safely and reliably deliver the required levels of service to customers and communities as it ages, necessitate a whole-of-life approach to decision-making. At the various key stages of an infrastructure's life from its early inception, through its design, construction, use and decommissioning (or repurposing), short and long-term considerations need to balance asset performance, the costs to achieve it and the risks should it fail. This principle is at the core of sustainable life-cycle asset management as defined in ISO 55000 (International Standards Organisation, 2014) and the International Infrastructure Management Manual (Institute of Public Works Engineering Australasia, 2015). Originally considered an engineering discipline focused mainly on asset maintenance, asset management is now defined and considered a whole-of-life business process aimed at sustaining the services that an organisation provides to its customers through the efficient use of its assets (International Standards Organisation, 2014; Institute of Public Works Engineering Australasia, 2015; Lafraia et al., 2013; PIANC, 2008). As such, considerations need to be given to the importance of decisions and particularly at the creation stage of an asset. The Water Environment Research Foundation (2011) illustrates this point by showing a decreasing opportunity to influence the whole-of-life cost of infrastructure as it ages. As discussed in previous chapters, different materials deteriorate in distinct ways and thus require specific inspection and maintenance procedures. As such, once the decision has been made to build an asset using reinforced concrete, for instance, it locks in the approaches to manage this asset while making it costly to significantly alter the design (e.g. change of construction materials) during the operational life.

Achieving the desired life-cycle asset management objectives requires all parties (asset owner, designer, constructor, manager and operator) to trust and own the decisions made at the various stages of the infrastructure's life. To do so necessitates a holistic approach through (Gabbedy et al., 2016):

• Defining a clear set of objectives, including performance requirements
• Articulating a process for asset life-cycle management and optimisation
• Engaging individuals and teams to contribute knowledge and skills to the decision-making process
• Collecting and managing quality and reliable information
• Using effective tools/technology solutions that enable the management of information in support of collaborative decision-making

As outlined in previous chapters, the interaction between an ageing asset and its environment can significantly impact its performance and its Service Life, hence, why there has been greater emphasis on durability management across the world. According to ISO 13823, durability is 'the capability of a structure or any component to satisfy, with planned maintenance, the design performance requirements over a specified period of time under the influence of the environmental actions, or as a result of a self-ageing process' (International Standards Organisation, 2008). Durability design is, thus, a process that links materials selection to the targeted levels of services for the assets being created; that is, the whole-of-life return on an asset can often be closely related to its durability characteristics (Blin et al., 2011; International Standards Organisation, 2001–2011). For instance, the type and rate of materials degradation in an operating environment can have a marked impact on the need, extent and timing (and thus costs) for monitoring and repairing an asset so it remains serviceable. As a practical example, specifying a steel coating system that requires regular inspections and minor maintenance might be acceptable from a whole-of-life perspective for an asset that can be safely and easily accessed, as well as not constantly operated. The durability requirements may be very different if access was, on the other hand, risky, expensive and any interruption to service could affect the service levels and revenue.

Durability design is underpinned by the identification, assessment and mitigation of risks associated with the failure of assets (or their components) within their specified Design Life resulting in early (and often costly) treatments. This approach can both enable the design and construction of durable, less maintenance-intensive assets (i.e. balancing the capital and operational budgets from the outset), as well as minimise the risk of overdesigning through potentially onerous and less sustainable material selection (e.g. selecting a higher than necessary grade of stainless steel).

A consistent, systematic and transparent durability management process being a key enabler to achieve whole-of-life objectives, it needs to be integrated across all phases of an asset's life cycle, as suggested as follows:

- Identifying key performance objectives and parameters at the planning stage that can be converted into specific quantitative design, construction and operational criteria
- Using those objectives and parameters to develop a Durability Management Plan that informs the design and outlines a collaborative and effective process to address durability-related risks in order to produce designs that comply with the Design Life requirements
- Applying the durability process into the construction phase, working hand in hand with the quality control and assurance process and non-conformance reporting to provide a more satisfactory basis for determining which option will achieve the original design intent, rather than relying on an 'on the spot' expert opinion
- Inspecting and maintaining the assets in line with the requirements of the design so that they achieve a Service Life equal or greater than the specified Design Life

As mentioned earlier, however, durability design can only help achieve optimised life-cycle objectives if it is adopted by the whole project team, supported by reliable information and deployed via a medium (tool/documentation) that is useable and effective. The following sections outline how this approach can be implemented during the planning, design, construction and operational phases of an infrastructure asset's life.

PLANNING PHASE

At this early stage of the asset creation process, the definition of the services required to be provided is essential to determine the need for a particular portfolio of assets. This is a very important phase, as several factors can impact the subsequent phases of the assets' life, for instance:

- The base environmental (e.g. soil and groundwater) conditions to which the asset will be in contact with and the associated durability risks
- The potential impact of operations onto the exposure conditions impacting durability, such as the production of brine in a desalination plant or elevation of acidic gases due to exhaust emissions in road traffic tunnels
- The balance of short-term (construction) and long-term (operations) considerations as a result of the type of contractual arrangement selected (design and construct, public private partnership, etc.)
- The possible funding/budget constraints that can lead to a focus on capital expenditure (CAPEX) minimisation resulting in likely higher operational expenditure (OPEX)

While at this stage the assets may not even be conceptually designed fully, their base form and functionality can be defined to support the preparation of cost estimates for the business case that ultimately drives the decision to proceed to asset creation. As a starting point, it is essential to articulate what function (or Level of Service) an asset needs to achieve over what Design Life. As mentioned earlier, this includes assessing the impact of an asset failure on the asset itself and on the broader network it supports in order to support the development and implementation of appropriate risk management strategies. Generally failure modes can include:

- Asset deterioration (expected and predictable asset/material degradation over time)
- Growth in demand to or that exceeds original asset design capability (including unexpected catastrophic events, changed regulatory requirements, etc.)
- Change in use that exceeds the original design capability (changing market conditions that require a shift in focus for a business and the use of its assets)
- Economic failure (through obsolescence, more cost-effective solutions for delivery of the same service, new disruptive technologies, etc.)
- Human error (including third-party damage, etc.)

The ageing of infrastructure is mostly concerned with the first mode of failure listed earlier.

The planning phase provides the opportunity to outline an overarching strategy for the life-cycle management of the assets and the associated risks that need to be managed in the design, construction and operations of the infrastructure. As the potential for innovation is still very high at this early phase, considerations should be given to sustainability and resilience principles to inform the subsequent phases on the appropriate level of future proofing to be designed and built into the assets.

DESIGN PHASE

Concept design

The concept design is typically produced as part of a reference design phase or similar (depending on the procurement method for asset creation). At this stage, the focus is on defining the key objectives, requirements and constraints of the asset creation process. As a result, the following founding documents for the life-cycle asset management process are typically prepared:

- A contract specifying the Design Life requirements, the design process (including the level of technical review and importance given to durability) and framework (articulating how the different parties will collaborate to achieve the contract objectives).
- Asset management guiding principles that can be articulated in a strategy and associated plan. This can state the objective to balance construction and operational costs, information requirements (particularly during commission/handover phases) and a 'code of maintenance standards' or similar.
- A Durability Management Plan outlining a preliminary material selection to achieve the required asset performance and key durability risk mitigation approaches.
- Preliminary designs incorporating functional (e.g. serviceability and availability) and technical quality (including durability) requirements.
- A life-cycle costing assessment taking into account the earlier to present the estimated CAPEX and OPEX for the project.

Design Life requirements

It is important to first define what life is required for the assets being created and what constitutes an end of life (based on the modes of failure). As previously discussed, the Service Life is intended to be, by definition, greater than the Design Life; thus, the figures provided for the latter in contractual documents are considered minima. If the definition of 'Design Life' adopted is aligned to the like of ISO 13823 (International Standards Organisation, 2008), and unless specified otherwise in overarching asset management documents, the infrastructure and the materials it is made of need to last the required life with (minor) maintenance (as opposed to major activities including repairs and partial replacements). Several international standards, namely ISO 15686.1 & 2 (International Standards Organisation, 2001–2011) and ISO 2394 (International Standards Organisation, 1998), can be used for guidance on the appropriate Design Life.

The breakdown of the asset (in an early hierarchy) needs to be commensurate to the asset performance objectives, including potential life-cycle cost implications, taking into account practical considerations. For instance, specifying a one hundred-year Design Life for all components of a bridge may not be achievable, or possibly not without a higher CAPEX than more 'standard' designs.

That said, the criticality of an asset and the trade-off considerations between CAPEX and OPEX over the life of the infrastructure could dictate design lives in excess of those specified in design standards. An example was the three hundred-year Design Life requirements for the Gateway Bridge (Brisbane) that required a nonstandard approach to the design (e.g. stainless steel reinforcement utilised in the pile caps) (Connal and Berndt, 2009).

Together with having well-defined Design Life requirements, it is useful for any asset creation project to set a maintenance strategy for the asset(s), as one can influence the other. For instance, knowing whether the infrastructure can or will be maintained (and whether this can be done preventatively or would have to be reactive), or the amount of deterioration that can be accepted before an intervention is triggered, can help define the Design Life.

Durability design methodology

Once the Design Life and the maintenance strategy are specified, the durability process can be formulated and implemented. It is important that the assumptions and inputs that underpin the durability design are clearly and transparently articulated to enable checking and verification of the processes and the outputs it generated.

It is the key for the overarching objective of a durability approach to have its process well understood by all parties such that it can be implemented consistently and effectively throughout the design. ISO 13823 (International Standards Organisation, 2008) and ISO 15686 (International Standards Organisation, 2001–2011) are key complementary international guides for durability design. They can be further supported by a number of documents including guidelines, codes and standards across the world and industry sectors (e.g. CEN, 2004;

Concrete Institute of Australia, 2014; CIRIA 1983; PIANC, 2008). ISO 13823 'specifies general principles and recommends procedures for the verification of the durability of structures subject to known or foreseeable environmental actions, including mechanical actions, causing material degradation leading to failure of performance' (International Standards Organisation, 2008). ISO 13823 outlines a limit-state approach to design for durability as follows:

- **Determine** environmental exposure, which is defined as 'external or internal influences (e.g. rain, UV, humidity and soil constituents) on a structure that can lead to an environmental action'
- **Identify** degradation mechanisms including:
  - Transfer mechanisms (e.g. direct exposure, condensation, diffusion, etc.)
  - Environmental actions (e.g. corrosion of metals, sulphate attack of concrete, weathering of timber and polymers)
- **Consider** limit states based on the action effects on a structural component as either:
  - An **ultimate limit state** when the resistance of the structure or its components become equal or greater than what it can withstand
  - A **serviceability limit state** when local damage or displacement affects the function or appearance of the structure or its components

Assessment of macro-environments

The first step is to assess the macro-environments, for example how aggressive the conditions are for the proposed assets and the materials they are potentially made of. This would typically include the following activities:

- Listing of the different types of macro-environments: soil(s), groundwater(s), seawater, liquids/chemicals contained/transported/contacted by the assets, atmospheric, gas(es), etc.
- Breaking down these environments into relevant durability exposure zones, typically in alignment to applicable standards and guiding documents. As an example, seawater exposure can be divided into the following zones, each with its own level of aggressiveness: below seabed, submerged, tidal, splash, spray zone.
- Identifying relevant/applicable guiding documents for the classification of the exposure conditions.

This is illustrated in the following Tables 6.1 and 6.2.

Table 6.1 Examples of environmental influences and agents as per ISO 13823

Locations/environments	Outside atmosphere	Outside – ground or water	Inside
Influences	Rain Air constituents, contaminants or pollutants Wind Temperature and humidity Sun	Water Soil constituents Spills/leaks	Humidity and temperature Contaminating materials Water and sewage Stored chemicals Activities causing wear
Agents (causing environmental action)	Liquid (rain, condensation) Gas (water vapour) Oxygen, carbon dioxide, ozone Acidic compounds from certain polluting sources – acid rain Oxides, particulates Airborne chlorides Freeze-thaw cycles UV radiation, IR radiation	Chlorides, Sulphates and other salts Acids (from decomposition of organics or ASS/PASS)	Liquid (condensation) Gas (water vapour) Oxygen, carbon dioxide Oxides, particulates

Source: International Standards Organisation. (2008). ISO 13823 General Principles on the Design of Structures for Durability. ISO, 39p.

Table 6.2 Examples of exposure classification based on relevant standards

Locations/environments	Steel and alloys	Reinforced concrete
Outside Atmosphere	B to AS 4312/2312 [Australian Standards, 2008; Australian Standards, 2014) (C2 to ISO 9223 [International Standards Organisation 2012a])	B1 to AS 3600 (50-yr Design Life) (Australian Standards, 2009a) and AS 5100.5 (100-year Design Life) (Australian Standards, 2017)
Outside – Ground or Water	Mild to AS 2159 (Australian Standards, 2009b) Possible ASS/PASS	B1 to AS 3600 (50-yr Design Life) (Australian Standards, 2009a) and AS 5100.5 (100-year Design Life) (Australian Standards, 2017) with the possible exception of the area of ASS/PASS
Interiors	A to AS 4312/2312 (C1 to ISO 9223 [International Standards Organisation, 2012a]) for air-conditioned buildings B to AS 4312/2312 (Australian Standards, 2008; Australian Standards, 2014) (C2 to ISO 9223 [International Standards Organisation, 2012a]) for non-air-conditioned buildings	A2 in non-residential fully enclosed building (except for a brief period of weather exposure during construction) B1 for industrial building subjected to repeated wetting and drying to AS 3600 (Australian Standards, 2009a)

Source: Standards Australia, AS 3600 Concrete Structures, SAI Global, Sydney, Australia, 198p, 2009; AS 2159 Piling – Design and Installation, SAI Global, Sydney, Australia, 97p, 2009; AS 2312 Guide to the Protection of Steel Structures against Atmospheric Corrosion by the Use of Protective Coatings, SAI Global, Sydney, Australia, 133p, 2014; AS 5100.5 Bridge Design–Concrete, SAI Global, Sydney, Australia, 229p, 2017.

Table 6.3 Examples of hierarch or breakdown of assets

Asset group	Asset category	Asset components
Facilities assets	External services	Drainage
		Fire
		Light and power
		Sewer
		Water supply
	External works	Car parks
		Equipment
		Fences and walls
		Landscaping
		Noise barriers
	Station	Building
		Platform
		Services
		Structures

Categorisation of asset elements – asset register

A preliminary asset register can be created based on an agreed hierarchy (breakdown of the infrastructure into components and subcomponents for instance) at this phase that helps identify which assets are in what environments, as those will influence the ageing process. Following is an example of an asset breakdown and a useful guide to assist in classifying assets is ISO 12006 (International Standards Organisation, 2015a and b) (Table 6.3).

Such an approach can also help identify micro-environments that can potentially require specific considerations, for example the localised part of an asset where salts may deposit but cannot be easily washed down by rain or inspected.

Assessment of deterioration mechanisms

The next step consists of identifying the ageing mechanisms for the infrastructure materials being considered. Assets can be exposed to multiple environments (if they have internal and external surfaces, for instance, or have elements exposed to different conditions, such as a bridge deck's top surface vs. its soffit). Within the same environment, they can also be subjected to multiple deterioration mechanisms: for example, the chlorides and sulphates in seawater can attack concrete reinforcement and the concrete matrix itself, respectively. It is, thus, important to identify the governing mechanisms (or that driving the 'critical path' of deterioration and thus most significantly impact the Design Life if not adequately designed for) and any compounding/worsening effects due to a combination of actions (e.g. early concrete cracking allowing faster penetration of chlorides) (Table 6.4).

Table 6.4 Example of deterioration mechanisms

Locations/ environments	Steel and alloys	Reinforced concrete	Polymers (includes protective coatings)
Outside Atmosphere	Corrosion leading to oxides/hydroxides buildup (rust) and eventually failure of element Possible galvanic corrosion when electrically connecting dissimilar metals Creation of crevices at bolted or partially welded connections can produce a micro-environment where moisture is retained and very rapid corrosion occurs	Reinforcement corrosion and concrete delamination resulting from carbonation	Chemical degradation leading to loss of strength, cracking Photo-oxidation (UV exposure) resulting in degradation of mechanical properties, discolouration, hazing, dullness
Outside – Ground or Water	Corrosion leading to oxides/hydroxides buildup (rust) and eventually failure of element Possible galvanic corrosion when electrically connecting dissimilar metals Stray current-induced corrosion	Degradation of the concrete matrix through sulphate attack/acid attack (e.g. ASS/PASS) or Possible exposure to soft waters Reinforcement corrosion and concrete delamination resulting from chloride attack Stray current-induced corrosion	Swelling and dissolution when exposed to liquids leading to a weakening of molecular bonds and a possible reduction reduced of glass transition temperature (becoming weak and rubbery)
Interiors of Assets	Corrosion leading to oxides/hydroxides buildup (rust) and eventually failure of element Possible galvanic corrosion when electrically connecting dissimilar metals	Reinforcement corrosion and concrete delamination resulting from carbonation	Chemical degradation leading to loss of strength, cracking

Durability risk assessment

By applying risk management principles (Standards Australia, 2009c), the durability risk would be the product of the likelihood of asset failure by its consequence. The likelihood of failure is related to the condition of the assets over time and, thus, how it ages within its Design Life. The consequence of failure relates to the impact on the levels of service the assets are required to provide. ISO 13823 (International Standards Organisation, 2008), for instance, proposes four categories ranging from minor and repairable damage without injuries to people (International Standards Organisation, 2014) to loss of human life or serious injuries or considerable economic, social or environmental consequences (PIANC, 2008). In addition to those, a fifth category that could be called 'negligible' is at times introduced to produce a symmetrical matrix that aligns with that presented in Table 6.5 of HB 436 (HB, 2004).

As discussed previously, the concept of risk management is fundamental to asset management and, thus, a key driver to durability design. As such, durability risks need to be identified as part of the process by assessing the likelihood of damage or failure of material/treatment options and understanding from the designer/constructor/operator/owner of its consequences. Such risks can be included in safety in design initiatives, as the potential impact of material selection on the future inspectors and maintainers of the assets need to be taken into account. Moreover, access considerations to perform inspection, maintenance, repair and/or replacement activities in the future can affect the consequence of asset failure and, thus, the durability design.

As part of the design process, when the durability risk is considered too high, it could be reduced by either one or a combination of the following:

Table 6.5 Example of durability risk matrix based on (HB 436, 2004; AS/NZS 4360, 2009c)

Likelihood of damage/failure	Consequence of failure/damage				
	Negligible	Minor	Moderate	Major	Severe
Almost certain	Medium	High	High	Very high	Very high
Likely	Medium	Medium	High	High	Very high
Possible	Low	Medium	High	High	High
Unlikely	Low	Low	Medium	Medium	High
Rare	Low	Low	Medium	Medium	High

- Decreasing the likelihood of failure by proposing a more durable option (e.g. more corrosion-resistant metal, greater concrete cover to reinforcement, use of cathodic protection, more robust coating system). However, this has to be balanced by taking into account constructability and cost implications (i.e. can it be built, procured or even afforded?).
- Putting in place mitigation strategies aimed at limiting the consequences of any damage/failure, for example, having sufficient redundancy in the process equipment, restricting access.
- Tailoring the inspection and maintenance plan (and especially the frequency of these activities) to detect early signs of unacceptable deterioration in order to proactively plan for repair and/or replacement. Hence the importance of identifying potential access challenges as mentioned earlier.

Performance criteria and future predictions

Based on the deterioration mechanisms identified, performance criteria can be articulated to ultimately achieve the required Design Life. These criteria can, according to Report 103 (PIANC, 2008), be broken down into two main categories:

- Functionality including prime functional requirements, serviceability, availability
- Technical quality, such as safety, security, social compatibility, environmental, aesthetic, durability, sustainability, constructability, inspectability, maintainability and reuse

The durability criteria can, for instance, be presented as a maximum allowable corrosion loss over the asset life or an acceptable time for the initiation of concrete reinforcement corrosion.

Once the performance criteria for each asset and its elements (as applicable) have been set, materials options can be considered and assessed, taking into account the durability risk(s). As outlined in ISO 13823 (International Standards Organisation, 2008), a number of approaches can be considered to predict the future performance of the material options, such as use of standards/codes/guidelines, historical data on asset performance, use of models and physical testing of adjacent/similar assets. An example of a standard-based approach is provided in Table 6.6. Note that while this can be well-suited for the specification of minimum durability requirements in macro-environments, a more tailored approach (such as modelling) may

Table 6.6 Example of minimum durability requirements for steel and concrete

Locations/environments	Metals	Reinforced concrete
Outside Atmosphere	Corrosion rates for steel in AS 2312/ISO 9223/ISO 9224 (Australian Standards 2014; International Standards Organisation 2012a, 2012b): B/C2 (first year): 1.3–25 μm/yr B/C2 (first 10 years): 0.5–5 μm/yr B/C2 (steady state): 0.1–1.5 μm/yr Corrosion rates for zinc in ISO 9223/ISO 9224 (International Standards Organisation 2012a, 2012b): C2 (first year): 0.1–0.7 μm/yr C2 (first 10 years): 0.1–0.5 μm/yr C2 (steady state): 0.05–0.5 μm/yr Corrosion rates for aluminium in ISO 9223/ISO 9224 (International Standards Organisation 2012a, 2012b): C2 (first 10 years): ≤0.025 μm/yr C2 (steady state): 0.01–0.02 μm/yr	50-year Design Life to AS 3600 (Australian Standards, 2009a): B1: covers of 40 mm (32 MPa), 30 mm (40 MPa) and 25 mm (≥50 MPa) Tolerances as defined in overarching specification 100-year Design Life to AS 5100.5 (Australian Standards, 2017): B1: covers of 45 mm (32 MPa), 40 mm (40 MPa) and 35 mm (≤50 MPa) Tolerances as per section 4.10.3.1 Maximum permitted crack width: 0.2 mm to VicRoads (2017) specification 610 (Table 610.241)

be required for micro-environments, for instance. The complexity and accuracy of models can vary between the modes of degradation they are aiming to simulate, and this needs to be considered when using such an approach. As an example, the diffusion of chloride into concrete to reach the reinforcement, posing a steel corrosion risk within the concrete, can be modelled deterministically using a relatively simple equation or probabilistically based on more complex relationships between parameters. The former approach may be useful for quick high level estimates of reinforcement cover requirement and/or life as a first pass, while the latter can be used to more robustly simulate a scenario with greater confidence in the outputs provided (Table 6.6).

Options analysis and decision-making

Using the durability design process (see previous section on durability design) can support 'optioneering' not only of different designs for the assets but also for the materials they are made of. A different approach can be used to evaluate and select options (IPWEA, 2015; Lafraia et al., 2013; PIANC, 2008), but it would focus first on the functional requirements before taking into account the technical considerations (see see previous section on performance criteria and future predictions). It would also take into account Whole of Life (WOL) considerations that can also assist with the development of the design phase, such as:

- The use of (in situ and ex situ) sensing technology for asset monitoring can support the risk management process, but the use of sensors need to be carefully balanced based on WOL considerations of the sensor system and associated data management.
- Minimising the risk of rework wherever practicable over the infrastructure's life could be achieved by combining life-cycle asset management principles, sustainability, resilience and durability considerations.
- The maintainability, including the accessibility and availability, of assets associated with adjacent rail infrastructure needs to be clearly specified in relation to durability design.
- The fewer the number of assets and lower the complexity of the assets (to build and maintain), the lower the life cycle costs.
- The type of assets (assuming that they are well-designed and constructed) can drive different maintenance costs: a bridge could have a much lower maintenances costs (in proportion to the asset's value) than a tunnel.
- Safe access to assets for inspection and maintenance activities can be a major factor in the costs of these activities and needs to be factored in.

- Should a concession-type contract be considered, the life cycle cost implications of the concession duration and requirements on the assets at handover (residual life, condition, performance) should be carefully evaluated.
- Data has inherent value for an asset, as it can support optimised decision-making; thus, allowances should be made for the seamless transition of data throughout the assets' life.

The approach for option selection can include a mix of quantitative (e.g. CAPEX and OPEX estimates) and qualitative measures (e.g. social, environmental, technical and reputational impacts that can be evaluated in a multicriteria-type assessment). In a similar manner, when a number of materials options are available, the risks associated with durability, constructability and qualitative cost consideration can be rated (International Standards Organisation, 2014; Institute of Public Works Engineering Australasia, 2015; Lafraia et al., 2013; PIANC 2008; Water Environment Research Foundation, 2011), adding the figures and selecting the one with the lowest risk value. It is worth mentioning that if no major maintenance is allowed to achieve the required Design Life, any option requiring more than minor 'refurbishment' should be eliminated.

Durability management plan

Such a document, or variation thereof, can be a useful way to document the durability process, the inputs fed into it and the outputs it generated. In some geographies/sectors, such a document is mandated with prescribed requirements on its content. Blin et al. (2011) proposed the following template for a Durability Management Plan (DMP), though this can be tailored to the complexity and risks associated with the infrastructure being designed.

1. Project overview and overall durability objective.
2. *Approach and scope:* Extent and methodology of the durability design.
3. *Referenced documents:* List of standards, codes and guides and project-specific documents upon which the durability design in the DMP is based.
4. *Terminology:* Definitions, abbreviations and symbols used in the DMP.
5. *General overview of assets:* Asset hierarchy and basic list of assets and subassets.
6. *Overview of environmental exposure categories for the assets:* Detailed information and pertinent test data could be provided in an appendix.
7. *Overview of deterioration mechanisms for construction materials in each exposure:* Again, details could be outlined in an appendix.

8. *Durability requirements:* Outcomes of the materials selection based on the assessment/prediction/modelling of future deterioration and the durability risk associated with material deterioration. More detailed information could be provided in an appendix.

Possible appendices:

- Details on environmental exposure classification
- Details on materials degradation mechanisms
- Detailed durability considerations and requirements
- Detailed durability risk assessments and mitigation options
- Predictive modelling/limit-state calculations
- Detailed scenario/'optioneering' considerations

Collaboration for effective outcomes

Best outcomes for the design can occur when the environment supports a culture of collaboration, high performance and innovation. This is where the future risks and opportunities associated with the use and management of the infrastructure as it ages can be discussed at the earliest stage of the asset creation process so it influences the decisions made. In that context, asset management/durability-related documents and outputs (e.g. Durability Management Plan, preliminary asset register) are efficiently used by other disciplines, including civil and structural designers, for instance, as well as cost estimators to produce an optimised concept design.

Detailed design

At this next stage of the asset creation process, the following documents can be produced to integrate whole-of-life/infrastructure ageing considerations into the design:

- An updated/more detailed DMP
- Durability reports or memos providing advice to discipline leaders/ designers for incorporation into design packages
- As-designed asset registers
- Technical specifications and/or procedures for the mitigation of specific corrosion risks

With a strong technical and engineering focus to design assets complying with the project scope requirements the asset creation process has to balance the following asset management risks:

- Minimising CAPEX such that it increases significantly the OPEX over the asset's life. An example could be underestimating the corrosion rate of an element that subsequently needs to be coated to achieve its

required life (which can also bring about safety risks associated with the coating and monitoring activities in an operational environment).
- Overspending (sometimes referred to as 'gold plating') on the CAPEX to minimise risks of asset failure below a manageable level. An example of overengineered solutions could include the use of stainless reinforcement in a high-quality concrete mix for an entire slab when only a small section of it would be exposed to very aggressive conditions (and could thus receive a tailored treatment such as a coating).

Overengineering could also increase the durability risk and lower the asset's life. For instance, too high a cementitious content in a reinforced concrete mix (on the incorrect assumption that more cement always provides greater durability) could negatively affect workability and placement, as well as cause cracks due to heat of hydration and temperature differentials. The cracks could have detrimental structural and durability impacts.

Early in this phase, the preliminary DMP may be revised to more comprehensively document exposure conditions (macro- and micro-environments), ageing mechanisms, durability risks and material selection process. At this stage, site testing (e.g. from soil and groundwater samples) specified in the preliminary DMP would be undertaken together with trials (e.g. concrete mix designs and test coating applications). The revised DMP would outline minimum durability requirements for each material in each distinct environmental condition in alignment with the selected maintenance strategy for the nominated Design Life. Design leads would first use the DMP as a basis for material selection and can seek more specific advice as required. As such, the DMP should be user friendly for the designers/engineers who use it as a basis of durability design, as well as for reviewer(s), contractor(s), operator(s), concession holder(s) and asset owner(s). The information from the DMP in terms of materials selection and workmanship requirements to achieve durable outcomes would flow through design packages that typically include drawings, specifications and design reports, as well as digital models (e.g. Building Information Modelling, 3D finite element representation). The design specifications would list Quality Assurance/Quality Control (QA/QC) requirements, including the type and frequency of testing and inspections, as well as hold points necessary to demonstrate compliance to the design including for durability parameters.

As the detailed design phase progresses, specific design queries may require the preparation of durability reports or memos that would apply the durability risk-based process documented in the DMP to provide the required technical advice. This includes, for instance, considerations on the suitability of alternative materials, proposed construction procedures and/or review of QA/QC procedures and associated test results (e.g. reviewing concrete test results including chloride diffusion coefficient measurement to release a hold pint prior to full production). For this process to be effective and lead to the provision of timely support, it can prove helpful to embed a durability team into the core design team. Critical to the successful development and implementation of the durability process, and associated impact on the project's life-cycle management objectives, is the buy-in not only of designers but also construction and Operations & Maintenance (O&M) leads. This can be achieved by seeking their inputs, feedback and comments on the durability design, the DMP and durability-focused documents, as well as the inclusion of durability team members in design (including safety in design) meetings. Equally important to the process is the quality of the input data and usability of the information outputs. As such, a plan can be developed to map the type of input data necessary to develop the durability requirements and the output or deliverable (including its timing) that had to be generated. The deliverable from the design phase would then form the input to the construction phase. Specific advice may be required in order to optimise the materials selection in high durability risk areas, for instance:

- Concrete mixes in acidic, high chloride, high sulphate and/or high magnesium environments
- Use of polymers (including fibre reinforced) as a durable alternative to metallic options (that can also be easier to transport and install)
- Tailoring of the grade stainless steel (e.g. local use of super duplex in high chloride environments)
- Use of separating nonconducting materials in connections between dissimilar metals to avoid galvanic corrosion (e.g. stainless steel fixings connected to coated plain steel pipes)

All the developing designs and material durability data should be captured in an evolving asset register that is progressively developed (based on the asset hierarchy structure produced in the preliminary design phase). The information contained in the register would depend on the specific contractual requirements from the asset owner. A list of key durability-related parameters could include the following: unique tag number, duty description, Design Life, materials of construction, environmental exposure, durability risk, durability issues (deterioration mechanisms), minimum durability requirements, location of asset, manufacturer (if relevant), drawing references, links to pertinent inspection and repair procedures, inspection and maintenance cycle frequencies and condition rating.

The inspection procedures nominated in the asset register can be developed in the detailed design phase to provide useful guidance and information from the onset of the operational life of the asset. This could include any safety issue (particularly if it relates to a risk associated with the deteriorating of the asset or access or testing equipment requirements), the performance/ condition monitoring (e.g. rating scale) process and the criteria triggering what repairs.

In terms of maintenance activities, the register would provide a schedule for minor maintenance (e.g. asset cleaning) but would leave the timing of any major repair or replacement to be driven by the overall maintenance strategy and the evolution of the assets' performance and condition.

Key durability risks may require the preparation of technical specifications or procedures such as for the selection, application and repair of coating systems, or the identification and mitigation of galvanic corrosion, for instance. Toward the end of the detailed design phase, the preliminary requirements for postconstruction inspection and repair procedures should be nominated. It can prove useful to develop construction repair procedures (such as for the treatment of concrete cracks), and then such documents are made available before they are needed, thus, mitigating the risk of associated delays during the construction.

CONSTRUCTION

It is common for major projects to have the construction phase start while design is still underway. As such, it is important that the durability process, which underpins the life-cycle optimisation objectives, is put in place early to support construction activities. An effective process typically includes:

- Having construction leads being familiar with the DMP in conjunction of their corresponding design packages
- A framework to respond to Requests for Information (RFIs) from the construction team, enabling the design team to review the contractor's submittals, as well as approve materials and methods
- Review of procurement packages against the requirements of the specification, which may include reviewing alternative material options
- Inspection and supervision of construction activities in application of the QA/QC plan to document compliance to the specification and identify nonconformances. A durability specialist is often called in to provide support for the following activities:
  - Concrete mix design modifications and delivery compliance
  - Major concrete pours
  - Coating applications
  - Review of repair methods (coatings, concrete) and their implementation so the specified design lives can still be achieved
  - Inspections at hold points including as-built assets
- Organisation of design/construction interface meetings, particularly for critical assets
- Provision of a commissioning reports and associated as-built information including an updated asset register

Due to the pace of construction activities, the development of a trusting relationship between the design, construction and O&M teams can be both critical and challenging, particularly when trying to balance CAPEX and OPEX considerations. Having construction leads, spending time at the design office and design leads mobilising to the site can allow for effective communication and collaboration. Due to the importance of material selection and asset integrity during the construction process, the presence of a durability specialist on site can prove to be effective in mitigating and addressing issues promptly, as well as building trust on the ground in the project's durability process.

The conformance of the built infrastructure to the design intent so it can achieve the required life can be challenged by the need to adapt standard quality systems and at times entrenched work practices. Life-cycle management and durability considerations can affect every aspect of the construction phase, including information and inputs into detailed specifications, requests for quotations, materials and equipment procurement, production control forms, work method statements, QA/QC procedures, Inspection and Test Plans (ITPs), material compliance certification, testing and trials, product and installation warranties, product and equipment commissioning procedures and O&M manuals. As there is rarely a single material solution to any given durability issue, this requires dynamic and effective interactions between the construction and design teams. The advantage of having outlined a clear durability process is that it allows for alternative materials or construction methods to be suggested so they can be evaluated against the criteria and use the process set out in the DMP. Such a transparent and consistent approach as to how alternatives are selected and/or rejected, along with the evaluation of the direct and indirect costs associated with maintenance and renewals of one option versus another, is likely to be considered robust enough for reviewers (e.g. acting on behalf of asset owners) to sign off on changes. As an example, the risk-based approach described previously can be used to assess materials options taking into account safety, environmental, durability, constructability and life cycle cost considerations. Should no major maintenance be allowed to achieve the required Design Life, any option requiring more than minor 'refurbishment' would be discarded.

The basis of design relies on the materials handled, placed and used on the construction site to achieve their required properties and performance. As such, it is essential that the outcomes of site trials, testing and proposed construction methods are adequately reviewed prior to full-scale implementation. It is equally important that the application of the approved procedures is assessed on site as per the QA/QC management plan. As shown previously (Aziz et al., 2011), a well-designed QA system is an important tool that, when implemented correctly, delivers benefits to both contractor and asset owner in terms of quality, time and cost. In addition to such system, having in place a 'no-blame' collaborative culture can also assist in the proactive identification and management of nonconformances.

While every effort can be made to achieve a build quality of the required standard, defects and nonconformances can still occur. These are typically identified through the implementation of the QA/QC plan and the required site inspections. In such instance, the impact of the defect on the durability of the affected asset needs to be assessed against the design criteria (and the risk of not achieving the required Design Life). Where necessary remediation measures may need to be designed, approved, implemented and checked on their performance. This is where the availability of pre-prepared repair procedures (with approved remediation methods and materials) can minimise the risk on the construction program as well life cycle costs. The design, construction and O&M teams in consultation with the asset owner can identify the short- and long-term risks of defect and non-conformances and, where possible, estimate the costs of asset failure and the required intervention(s) to assess if an acceptable balance of CAPEX versus OPEX can be achieved. Such repair procedures can also be used in the operational phase not only to support the record (made in the asset register) of repairs but also to be used for any subsequent need to remediate assets. Undertaking repairs may impact on the frequency of maintenance and long-term (localised) life of an asset.

The asset register can be updated progressively with as-built information incorporating any changes to the durability of the asset item resulting from change of materials or damage and repair of construction defects. The register can also refer to postconstruction inspection and repair procedures that can be established during the construction phase (which are tailored to the actual materials used for the as-built assets). The nominated inspection and (minor) maintenance (e.g. cleaning) frequency should also be clearly nominated in the asset register. Prior to asset handover to the O&M team, the as-built asset register can be uploaded onto a Computerised Maintenance Management System (CMMS), allowing it to be tested before being used during the operations phase.

In most projects, a commissioning and handover report is prepared to confirm compliance of the built infrastructure with the design requirements, close out any outstanding items (such as rectification works) and list all the documents that can be used for the future management of the assets (e.g. drawings, specifications, digital models and tools and as-built asset register).

OPERATIONS

The operational phase spans what is at times called the 'beneficial/productive life' of the assets. This is indeed the period where the infrastructure gets to deliver the levels of service that it was intended to provide to the asset owner's/manager's customers. This is also a time where the whole of life objectives set at the asset creation phase can be challenged through, for instance, decisions and/or events that happened earlier (e.g. defects, less durable material selection, change in maintenance strategy) and/or environmental changes (e.g. less demand than forecast for the services). The continued implementation of an asset management business process drives the regular assessment of demand, asset performance/condition and opportunities for improvements in order to balance the levels of service with risk and cost. In the O&M phase, durability information is, thus, needed to assess the current condition of the asset, estimate its remaining life and predict its long-term performance.

Key events relating to the life-cycle optimisation process include:

- Handover of construction documentation (including drawings, specifications, reports, digital models and the as-built asset register).
- Review of construction nonconformances and their resolution in terms of Service Life implications, including the type and frequency of inspections and maintenance activities and costs.
- For a new organisation managing the newly created assets, full development of the asset management system (AMS), including policy, strategic asset management plan, asset class plans, asset management information system or AMIS (International Standards Organisation, 2014). For an existing organisation having added asset(s) to its portfolio, this could trigger changes to the AMS to reflect the new capability/ services offered and the risk profile.
- Review, update and implementation of asset integrity inspections and monitoring activities.
- Implementation of minor maintenance activities.
- Optimisation of the maintenance strategies and schedules, as well as renewal planning.
- Delivery of major maintenance activities (triggered by condition/risk monitoring or possibly reactively, such as in emergency type scenarios) and/or renewal that aim to achieve the whole of life objectives.
- Continually review performance of the AMS and all its components to improve.
- Assessment of the need to decommission/dispose of assets.

Key enablers of the success of the AMS and the organisation's asset management objectives are the asset management processes described earlier, as well as data, people and tools/technology.

During this phase, it can be challenging for asset managers to assess whether ageing is progressing at the designed pace or whether it is slower (resulting in a Service Life likely well exceeding the Design Life) or faster (triggering more/earlier maintenance activities, repairs and/or asset renewal than originally anticipated). This illustrates the importance of information management throughout the life cycle of an asset. The collection of data (e.g. from condition assessments), its storage, its analysis so it becomes information and ultimately organisational knowledge that supports optimised decision-making is, thus, essential. For this reason, data is more and more identified and managed as an asset in itself. The early investment in an as-designed asset register, that subsequently becomes an as-built register, provides future asset managers with an invaluable record and baseline to allow for informed decisions.

The continual focus on optimising the asset life cycle drives the tailoring of the asset management team and the level of competency and skills required to achieve the organisation's objectives. It also enables the identification of external resources (e.g. specialist contractors, consulting partners) required to undertake specific tasks through the life of the asset.

To optimise decision-making during the O&M phase, particularly with a large infrastructure asset base, typically requires the effective use of tools/technology that manages data and provides the right information to the team members that need it (Gabbedy et al., 2016). For instance, the asset register usually lives in a Computerised Maintenance Management System (CMMS) that can be used to manage the work order and interface with an analytics package aimed at predicting future life scenarios. Enablers of successful ongoing optimisation using such tools typically include:

- Initial configuration so the IT solutions meet the needs and requirements of the organisation; for example, it enables the effective capture and reporting of maintenance history.
- Ongoing population of the CMMS with all relevant information and data (e.g. condition assessment findings) to the required standards.
- Integration across IT systems, for example: CMMS, operational systems, document management software, analytics package(s) and financial management applications.
- Training of personnel so they use the tools at their disposal to their full potential.

Whenever possible, the information collected during the O&M phase, if shared, can be analysed to help improve the future design, construction and maintenance of infrastructure assets.

DECOMMISSIONING/DISPOSAL/REUSE

As the infrastructure ages, asset managers may consider that the performance, risks and costs can no longer be adequately balanced (e.g. levels of service not being able to be met without significant capital reinvestment, risk of failure becoming too great). In the absence of a business case to extend the Service Life of the assets, the asset could be planned for decommissioning and potential disposal taking into the social, environmental and financial impacts of this decision. The potential reuse of infrastructure assets (given their footprint and location) is another consideration, whereby they could be adapted to provide different services to what they were designed for (e.g. the New York 'High Line' elevated linear park on a disused viaduct section). The potential future use of assets can, thus, be explored at the planning phase of new assets when considering the sustainability and cross-generational benefits that could be brought by such infrastructure.

CONCLUSION

This chapter discusses how a life-cycle infrastructure management approach, underpinned by a durability-based framework, can be adopted to deliver successful short- to long-term outcomes. A key focus on material/asset selection, together with an understanding of their life cycle costs and the prediction of their future performance enable the optimisation of the whole-of-life investment and help balance CAPEX and OPEX. This can, in turn, drive how the infrastructure's condition is monitored and improve the understanding of their modes of failure, as well as help use/develop tools and processes to better predict and manage future performance. This would also assist in improving the management of asset information, how to identify, collect and use the data that best support the decision-making processes.

A challenge and opportunity is, thus, to synergistically blend organisational governance, knowledge, processes, tools and resources to make the best of the ageing of infrastructure to ultimately provide sustainable outcomes for organisations and societies alike.

Key activities throughout the life cycle of an asset are summarised as follows:

1. Planning phase:
   - Assess base environmental conditions on asset life and performance
   - Consider life cycle risks and costs to achieve value and sustainability outcomes

2. Design phase:
   - Undertake durability assessment (Design Life, environment, deterioration mechanisms) and materials selection
   - Design access to allow for inspection, maintenance and repair based on constructability considerations
   - Prepare life cycle cost/assessment, if necessary, revise the design
   - Prepare plans for inspection, maintenance, repair, replacement and quality control during construction

3. Construction phase:
   - Review design and incorporate acceptable changes, which need to then be inspected and approved
   - Review procurement
   - Mitigate the risk of damage to assets during construction

4. Maintenance and operation:
- Ensure that the environmental exposure does not adversely change during the Design Life
- Implement inspection and maintenance plan (including cleaning, repair, replacement and monitoring)
- If damages/defects identified determine cause, record to provide feedback for future practice
- Manage assets to best balance levels of service, risks and costs

5. Decommissioning/Disposal/Reuse
- Assess social, environmental and financial impacts of this decision
- Implement plan to achieve desired objectives

REFERENCES

Aziz, A., Blin, F., and Dacre, M. (2011). Extension of asset life for Melbourne's Swanson dock, *Australian Journal of Civil Engineering*, 9(1): 35–46.

Blin, F., Furman, S., and Mendes, A. (2011). Durability design of infrastructure assets— Working towards a uniform approach, *Proceedings of the 18th International Corrosion Congress*, 20–24 November 2011, Perth, Australia, pp. 592–603.

CIRIA. (1983). Material durability in aggressive ground, Report R 98. Construction Industry Research and Information Association, 60p.

Concrete Institute of Australia. (2014). Recommended practice Z7/01durability planning, Concrete Durability Series, Concrete Institute of Australia, 73p.

Connal, J. and Berndt M. (2009). Sustainable bridges–300-year design life for the second gateway bridge, Brisbane, *Proceedings 7th Austroads Bridge Conference: Bridges Linking Communities*, 26–29 May 2009, Auckland, New Zealand, 8p.

CEN (2004). BS/EN 1992 Eurocode 2, Design of Concrete Structures, European Committee for Standardization, 230p.

Gabbedy, J., Tauvry, J., Blin, F., Furman, S., and Penson, D. (2016). Lifecycle asset management optimisation of desalination plant, *Proceedings of OzWater 2016*, 10–12 May 2016, Melbourne, Australia, 8p.

Institute of Public Works Engineering Australasia. (2015). *International Infrastructure Management Manual (IIMM)*, 5th ed. Sydney, Australia: IPWEA.

International Standards Organisation. (1998). *ISO 2394 General Principles on Reliability for Structures*. Geneva, Switzerland: ISO, 73p.

International Standards Organisation. (2001–2011). *ISO 15686 Buildings and Constructed Assets — Service Life Planning, Parts 1–9*. Geneva, Switzerland: ISO.

International Standards Organisation. (2008). *ISO 13823 General Principles on the Design of Structures for Durability*. Geneva, Switzerland: ISO, 39p.

International Standards Organisation. (2012a). *ISO 9223 Corrosion of Metals and Alloys–Corrosivity of Atmospheres–Classification, Determination and Estimation*. Geneva, Switzerland: ISO, 24p.

International Standards Organisation. (2012b). *ISO 9224 Corrosion of Metals and Alloys. Corrosivity of Atmospheres. Guiding Values for the Corrosivity Categories*. Geneva, Switzerland: ISO, 13p.

International Standards Organisation. (2014). *ISO 55000 Standards for Asset Management*. Geneva, Switzerland: ISO, 14p.

International Standards Organisation. (2015a). *ISO 2394 General Principles on Reliability for Structures*. Geneva, Switzerland: ISO, 123p.

International Standards Organisation. (2015b). *ISO 12006 Building Construction–Organization of Information about Construction Works—Part 2: Framework for Classification*. Geneva, Switzerland: ISO, 32p.

Lafraia, J., Hardwick, J., Berenyi, M., and Anderson, D. (2013). *Living Asset Management, Engineers Media*, 180p.

PIANC. (2008). *Life Cycle Management of Port Structures, Recommended Practice for Implementation—Report No. 103*. World Association for Waterborne Transport Infrastructure, 56p.

SA/SNZ HB 436:2013, Risk management guidelines. Companion to AS/NZS ISO 31000:2009.

Standards Australia. (2008). *AS 4312 Atmospheric Corrosivity Zones in Australia*. Sydney, Australia: SAI Global, 30p.

Standards Australia. (2009a). *AS 3600 Concrete Structures*. Sydney, Australia: SAI Global, 198p.

Standards Australia. (2009b). *AS 2159 Piling—Design and Installation*. Sydney, Australia: SAI Global, 97p.

Standards Australia. (2009c). *AS/NZS ISO 31000 Risk Management—Principles and Guidelines*. Sydney, Australia: SAI Global, 37p.

Standards Australia. (2014). *AS 2312 Guide to the Protection of Steel Structures against Atmospheric Corrosion by the Use of Protective Coatings*. Sydney, Australia: SAI Global, 133p.

Standards Australia. (2017). *AS 5100.5 Bridge Design—Concrete*. Sydney, Australia: SAI Global, 229p.

VicRoads. (2017). Specification 610, structural concrete, VicRoads, 50p. https://www. scribd.com/document/348933813/VicRoad-Standard-Specification-Sec610-April-2017.

Water Environment Research Foundation. (2011). http://simple.werf.org.

Chapter 7

Health monitoring and intervention strategies

'Not everything that can be counted counts, and not everything that counts can be counted' – Albert Einstein, Physicist

INTRODUCTION

Health monitoring, as well as the development and implementation of intervention strategies based on the information obtained, are core activities for the management of ageing assets. To start with, it is important to agree on the rationale for assessing the health of assets, typically referred to, from a technical perspective, as their performance and condition. This is because collecting, analysing and managing information on assets over their Service Life can consume significant cost, time and resources. Data is increasingly more often considered an asset in itself, and thus the industry focus seems to be shifting from obtaining more and more data (with a large amount not being used) to the 'right' data that can be turned into information, knowledge and ultimately underpin the 'right' decisions. Like other assets, and as discussed in the previous chapter, the 'right' data is that for which its performance (e.g. value and use) is best balanced with the cost of managing (including collecting) it and the associated risks (e.g. of not having it, or the data not being reliable). This leads to the importance of clearly understanding the need for what data is gathered in the first place. A strategic approach involves specifying data requirements by working backward from the decision(s) that needs to be made and what would enable them to be as optimised as they need to be in order to meet the organisational and asset management objectives.

The terms 'performance' and 'condition' seem to be, at times, used interchangeably, but this can cause some confusion. The International Infrastructure Management Manual (IPWEA NAMS, 2015) relates performance to the ability to achieve the required levels of service of an asset, whereas condition refers to its physical state. In that sense, the condition is an enabler of infrastructure performance (e.g. a bridge whose piers are in very poor condition may be at risk for not being able to carry the required service loads), which may lead to restrictions being put in place that impact road users. Asset Management: An Anatomy (IAM, 2015) states that 'the term asset health is used in relation to measures that monitor the current (or predicted) condition or capability of an asset to perform its desired function, by considering potential modes of failure'. This document differentiates 'lagging performance indicators' of health, such as asset failures or incidents, and 'leading performance indicators', such as condition (IAM, 2015). The former allows for lessons to be learnt from historical issues that can help refine strategies and practices, while the latter assist in predicting the future state of an asset (e.g. its rate of deterioration or ageing).

In light of the earlier, the key objective of health monitoring is, thus, to obtain information on performance of the assets during their life and on their ability to provide the required levels of service so that decisions can be made for their effective and efficient management. As an example, AS ISO 13822 links the assessment of structures with potential interventions (ISO, 2005), such as continuing the inspection and maintenance program, modifying it or intervening to repair, renew, upgrade, replace or decommission the asset. The process of choosing between these different intervention options to develop strategies and asset management plans is at the core of life cycle asset management. It typically requires consideration of the current and future/predicted levels of service, condition, costs (operational and capital) and risks (defined as the likelihood of asset failure, thus, linked to asset condition, multiplied by the consequence of such failure, often estimated as part of asset criticality assessments). As the time horizon depends on the types of consideration – for example, operational within twelve months, tactical within one to five years, strategic greater than years – the possible combination/permutation of options can become complex. The need to undertake and demonstrate optimised decision-making has led to the increasing use of software solutions of varying complexity (from basic formulas in tools such as MS Excel to applications using artificial intelligence). It is important to note, however, that the software is a tool and, at least at the time of this writing, cannot provide solutions by itself, thus requiring human understanding, experience and skills to outline what the optimum intervention strategy is.

The following sections present different examples of infrastructure health monitoring approaches, the rationale behind their use and how this links to the future management of the assets.

SITE SURVEYS, TESTING AND MONITORING

The performance and condition of infrastructure assets are typically assessed through regular and/or ad-hoc monitoring activities on site that can include checks, surveys and testing. For performance and condition monitoring that enables decision-making on future interventions, it needs to be developed to match the assets' mode(s) of failure.

As mentioned previously, performance monitoring focuses on obtaining data relating to the levels of service provided by the assets, and thus help assess the likelihood of not being able to deliver the required customer outcomes (e.g. a water pipe or reservoir may not be able to supply residents the amount of water that is needed). Performance monitoring generally involves the use of devices, such as meters, sensors and cameras that can provide information on the service(s) provided by the assets. The selection of which types of devices, the frequency of recording and their locations and numbers would be related to the risk of performance failure on the organisation managing the assets. For instance, the number of vehicles using a toll road would be expected to be a critical performance indicator for the road operator, as significantly lower volume than anticipated could have dire financial impacts. As the consequence of failure is high in that example, it is therefore important to be able to accurately assess its likelihood to thus enable risk-based decision-making. This would drive the placement of reliable sensors at key locations that provide data that can be trended over time in order to forecast future estimates.

The same principles apply for condition monitoring in that it aims to provide an assessment of the likelihood of asset failure, the difference being that it is now focused on the technical (rather than customer) levels of service or, put another way, the physical state of the infrastructure. The selection of condition monitoring techniques would also depend on the strategic needs of the organisation that manages the assets (e.g. to present a ten- to twenty-year investment program outlining all planned capital and operational expenditure) as well as its tactical (e.g. to provide key inputs into asset management plans and three- to five-year budget forecasts) and operational requirements (e.g. for an annual state of the assets compliance report, defect rectification, prioritisation of works within the annual budget).

There are many documents that outline the different ways the condition of assets can be monitored (IPWEA/NAMS, 2016; AS/ISO 13822, 2005; Ports Australia, 2014; Bertolini et al., 2013; Concrete Institute of Australia, 2015; CIRIA 2006, 2010), but broadly speaking these can be categorised as per Table 7.1.

The selection of condition monitoring technique(s) needs to be tailored to the failure mode(s). There are often several of them, and they can differ for various components of the assets. As such, it is useful to map/draw the failure pathways followed by the infrastructure as it ages to articulate what mode of failure will likely dominate, the types and number of events that will occur prior to reaching an unacceptable condition level, so that techniques can be selected to proactively monitor asset health over time (AECOM, 2012).

Table 7.1 Broad categorisation of condition monitoring approaches and examples

	Non-visual	Visual
Non-contact	For example, sonar to map the underwater sections of wharves, laser mapping of bridge elements above ground/water	Human observations from ground/platform, cameras mounted on objects/machines (e.g. helmets, rods, crawlers, robots, drones)
Contact	For example, resistivity testing of concrete, corrosion rate measurements, concrete delamination testing	Scraping and observation of surfaces (e.g. concrete, coatings)

Figure 7.1 presents a basic example of a failure pathway diagram for a reinforced concrete element in contact with seawater (e.g. reinforced concrete pile supporting a wharf deck). The earlier the likelihood and extent of deterioration can be estimated the greater the ability to plan (the impact on the selection of future interventions is discussed later on). As such, undertaking a visual survey of the concrete can only reveal delamination, spalling and atmospheric corrosion, and advanced signs of ageing. As repairing damaged concrete can be quite expensive, it can thus be more useful to select condition monitoring techniques that identify earlier stages of deterioration. That said, the events outlined in the failure pathway shown in Figure 7.1 do not follow each other in a linear timescale. For instance, the reinforced concrete piles, if designed, constructed and placed adequately, can appear to be in what would be referred as good to fair condition for, say, 90 percent of their Design Life (that could have been one hundred years), while it could take further five to ten years for delamination, followed by spalling (or detachment) from the structure. This is where knowledge of the assets (and thus having access to reliable information) and how they age is important. Being able to form a hypothesis on where the infrastructure condition is likely to be after a number of years in service, that is at what stage of the ageing process, is a complementary consideration for the selection of condition monitoring technique(s). Using the example presented in Figure 7.1:

- If the asset was, say, within 50 percent of its Design Life, and the (as-built) information available suggested that its durability should be as-designed, then it would be unlikely for chlorides to have triggered the initiation of corrosion yet. As such, a basic visual survey may be all that was needed to check that there is no evidence of early deterioration.
- If the asset's age was close to, say, 70–80 percent of its Design Life and no information were available on its design/construction (with experience suggesting lower durability than for modern standards), it would be possible for reinforcement corrosion to have initiated without the

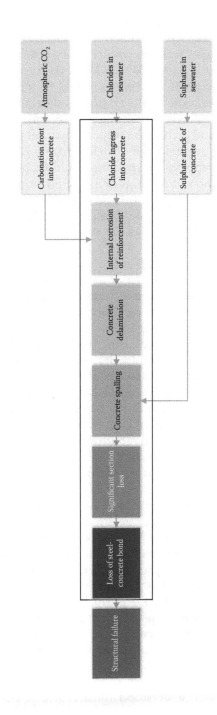

Figure 7.1 Example of failure pathway for reinforced concrete element exposed to tidal/splashing seawater (the primary deterioration mechanism has been outlined).

concrete showing any external signs of deterioration. In that instance, it would be helpful to assess the likelihood and possible rate of corrosion using techniques such as half-cell potential, resistivity and Linear Polarisation testing, as well as measurements of the depth of chloride ingress (Bertolini et al., 2013; WSCAM, Ports Australia, 2014).

• If spalling was observed, these areas would be measured (to record any increases in deterioration over time and/or plan for repairs for instance) and so would the remaining reinforcement bar diameter (to assess the current and predict the future structural risk). The concrete areas not showing any evidence of delamination (typically identified through hammer tapping) or spalling would be best assessed using the same combination of techniques as the previous example; this is to estimate the risk of significant increase of concrete deterioration within the short term, which could impact the selection of a type and timing of intervention, as well as the associated life cycle costs.

The aforementioned examples illustrate the interplay between the theoretical pattern of ageing (often represented as deterioration versus time curves), which can be drawn based on the durability design of the asset and the applied maintenance regime, and the actual condition observed on site. This is why condition can be considered a leading indicator of asset health: the comparison between actual against expected (using the baseline curve) allows for a recalibration and the prediction of future deterioration. This approach has its limitations, however, as will be discussed in the following section.

These examples show that likelihood of failure (in that instance associated with the expected physical condition based on the age of the asset, its primary materials of construction and the exposure environment) can be a driver for the selection of a condition monitoring regime. This could include the types of surveillance techniques but also the timing and frequency of their use. Health monitoring could be within the first part of the asset's life, involving mostly visual survey every five years or more before more advanced techniques are introduced as the infrastructure ages. The examples illustrate that monitoring asset health is typically achieved through a combination of visual and what is often referred to as detailed testing. The latter often includes a combination of nondestructive and destructive techniques that aim to provide specific information relating to the failure mode so as to gain a greater understanding of the current state of the asset than what can be visually observed (e.g. thickness of a protective coating, hardness of a polymeric material, pit depth of a steel element or presence of teredo worms within timber elements). There are many documents in the literature that list testing techniques available for the assessment of infrastructure assets (WERF, 2007; VicRoads, 2010; PIANC, 1990; Bertolini et al., 2013), most of those having been used for an extended period of time and thus having been widely understood and accepted across industries and geographies.

At the time of this writing, the use of remote monitoring did not appear to be a common method to record the condition of our infrastructure, as it was to assess asset performance. Across infrastructure sectors, sensors are in place to measure output ranging from traffic numbers on roads, on-time running of train/tram, water levels in tanks, flow rate in pipelines, number of ships/containers per berth/port to volume of resources loaded/unloaded from vessels. To the authors' knowledge, the use of remote condition monitoring of 'hard' infrastructure assets (as opposed to equipment, for instance, rail carriages) has so far been limited to examples such as:

• Video recording of assets, or parts thereof, because of their remoteness and/or criticality. This would mostly aim in monitoring any damage (or attempted damage) to an asset and not be used for close monitoring of asset health.
• The recording of movements by gauges (e.g. changes in crack widths, displacement) to identify trends in order to asses structural and durability risk, as well as tailor a scope of repairs, such as selecting flexible crack injection material for cracks subject to seasonal movements.
• The monitoring of the performance of cathodic protection systems that allow for the condition of the steel/reinforced concrete assets to be maintained/estimated.

This situation however, is changing and more 'smart' devices are becoming available in the market, supported by the development of the 'Internet of Things' (IoT).

The advancement of technology in general (including in electronics, software and telecommunication) and its increased affordability, combined with a need for safer and more efficient (in terms of optimising time, cost and quality) inspections have led to the trial and adoption of machines in place of humans. This has included the increased use of Unmanned Aerial Vehicles (UAVs), for instance: from nanodrones for use in small spaces to multirotor machines for façade inspections to larger aircrafts that can scan above-ground pipelines. While drones are most often equipped with high-resolution cameras, some can allow for the tailoring of the payload to different applications (e.g. use of infrared technology, laser scanning). Underwater inspections also increasingly include the use of Remotely Operated Vehicles (ROVs) that can allow for the safer collection of visual information than a diver; for example, the ROV does need decompression stops or can be operated in closer proximity to a berthed ship. ROVs have also been equipped with arms allowing them to successfully clean (e.g. using hydroblasting) and measure (e.g. via Ultrasonic Thickness Testing or UTT) the thickness of steel. The use of other machines has included robotic crawlers for vertical pipe inspection, remotely operated floating devices/boats equipped with multisensing cameras (e.g. sonar within fluids and laser in atmosphere), smart pigs for pipeline surveys, etc. While at present most of the robotic technology deployed on site for inspection requires tethering and/or line of sight with a pilot, in some instances semi-autonomous to autonomous vehicles are being used. This can include, for example, the use of a drone programmed to obtain footage of an asset along a defined flight path regularly within an industrial site. With the advent of fully autonomous vehicles on public roads within the next five to ten years, it is not far-fetched to imagine such machines being used to undertake surveys of assets, such as pavements, bridges and safety barriers, as well as possibly being deployed on rails and/or being used as part of a first response to an emergency situation (for engineers to assess the likely severity of damage and tailor the next steps accordingly).

In addition, pattern recognition software applications (some using artificial intelligence) are being increasingly used to extract information from site data, for example, processing numerous high-resolution pictures to report on the number and density of defects along an asset's surface and presenting associated infrared heat maps (geographical visualisation of the condition ratings using different colours from, say, green for very good condition to red for very poor/failed) to better support decision-making. In the medium to long term, it is possible that autonomous vehicles undertaking site survey activities could be equipped with deep learning type algorithms (as opposed to following a programmed route and task setup) to enable them to tailor the scope of their work to the data that has been collected on the asset to date.

Condition assessment regimes, and in particular with the increasing use of more advanced technologies, allow for the collection of a greater amount of data than ever before. However, greater data does not necessarily (and not often) translate into greater information and increased knowledge. This can indeed mean more 'noise', making it harder for humans to make decisions and thus possibly causing delays and inefficiencies, which negatively impacts the business case for the collection of such data. This is why there needs to be a strategic approach to data collection, analysis and management that aims to obtain the 'right' data (in terms of usefulness for decision-making as well as accuracy, quality and reliability) at the 'right' time and for which the most appropriate inspection regime and tools can be selected.

As briefly stated earlier, condition-based monitoring can, however, have limitations. The assessment of an asset's condition, for instance, is highly dependent on the definition of each rating in a condition scale. Different organisations use different scales with their own condition descriptions in an attempt to minimise subjectivity and support more quantitative analysis (e.g. 1–4 condition rating for VicRoads (2014), and the very common 1–5 as in the IIMM (IPWEA NAMS, 2015) and 1–7 in WSCAM (Ports Australia, 2014)). There are also differing approaches and levels of complexity, such as assigning a single rating to an entire element (e.g. condition 2 out of 5) versus providing a percentage breakdown of conditions within the built element (5 percent in condition 1, 25 percent in condition 2,

60 percent in condition 3, etc.). Some rating systems require the input of observed defect types, sizes, etc. into formulas to compute a condition rating. More detailed approaches are often used in order to estimate a representative condition rating for an asset that is not deteriorating in a uniform way (such as pitting corrosion on metallic surfaces, localised coating damage, or chloride-induced corrosion of reinforced concrete). It is also used for assets whose surfaces are exposed to different environmental zones (e.g. a steel pile can be in contact with the ground or seawater, including its low water zone with a potential increase risk of microbiologically influenced corrosion, tidal/splash or atmospheric).

The selection of an approach based on the likelihood of failure considerations alone is also not advisable, as the consequence of such failure must be taken into account as well to allow it to be risk based. As mentioned in the previously chapter, risk according to AS/NZS ISO 31000 is a product of the likelihood and consequence of an event (AS/NZS ISO, 2009). In the context of asset health monitoring, condition can provide the likelihood of failure and the consequence being assessed separately through a criticality analysis, for instance. Articulating the consequence requires understanding the mode of failure of an asset and the impact that this would have on its performance. A criticality analysis would typically encompass political, economic, social/safety, technological/technical, legal and environmental (often referred to as a *PESTLE* analysis) considerations with associated weightings to categorise and rank the assets based on their importance to the organisation's services. At an asset level, the same principles can be applied to outline the respective importance of asset components and their elements/parts (at this level the ranking is likely to be based on technical information only). This approach enables the risk to be rated, which can then be used to tailor the monitoring regime. As an example, for a wharf with a concrete deck supported on timber piles, not all its components have a wharf have the same level of structural redundancy/required reliability, and not all areas have the same capacity. The amount of timber that can be lost through ageing at different heights/locations along the piles can vary significantly (e.g. the shear connection to the deck/beam and that of the maximum bending moment could have greater impact on the structural behaviour than other locations). Because the poor condition of a critical area yields a higher risk that the same condition in a less important element/section, a different level of understanding (and possibly more detailed testing) could be required, particularly to help estimate the future rate of deterioration as a key input to decisions on possible intervention types and timing. A risk-based monitoring program could thus look like:

- High-risk assets/components: yearly inspections using a combination of visual and detailed testing
- Medium-risk assets/sections: visual inspection every two to three years with detailed testing triggered by the survey findings
- Low-risk assets/sections: visual inspection every five years

Table 7.2 Examples articulating the difference between a condition-only and risk-based approach for exposed timber piles supporting a marine jetty

	Condition/likelihood based	Risk based
Defined scope	All 300 piles (or a defined and deemed statistically representative percentage thereof) inspected in the submerged, tidal, splash/spray and atmospheric zones requiring significant resources (personnel, materials, budget) to complete	100 piles assessed at the tidal zone (high risk), 60 piles above water and 60 piles mid-water level (medium risk), opportunistic recording of low-risk sections/piles while undertaking assessment Higher value (high usefulness of data for lower collection cost)
Constrained program/budget	Prioritisation of data collection based on geographical and financial considerations, with important data being possibly missed, and difficulty for any extrapolation to reliably estimate future deterioration	Focus on high-risk piles first before medium-risk ones to obtain data that can be used to assess the need for, type and extent of, as well as timing of likely future interventions

Table 7.2 illustrates the difference between a condition/likelihood-based and a risk-based approach under two key scenarios:

- Where the scope is defined in terms of minimum requirements (e.g. all asset components to be rated on the one hand, all key risks recorded on the other)
- Where the budget and/or program is limited

In this instance, the asset could be a long marine jetty with over three hundred timber piles to be inspected both below and above water for the purpose of providing a condition assessment report (Table 7.2).

SAMPLING AND LABORATORY TESTING

As mentioned in the previous section, there are instances where the level of risk (likelihood of asset/condition failure versus its consequence) justifies the need for sampling and laboratory testing. This is usually the case where the materials characteristics at the time of design, construction (as-built) and/or currently cannot be easily estimated, that is, where uncertainties have an impact on the risk and could negatively affect decision-making.

This could take the form of collecting representative samples from areas (selection based on the risk assessment) for analysis on site. Examples can include:

- Removing an area of protective coating to identify its potential composition
- Extracting steel samples to assess their composition and characteristics, such as tensile strength and fatigue resistance
- Obtaining timber samples to identify species and structural parameters, such as strength and susceptibility to attack from (e.g. marine borers)
- Retrieving concrete cores to assess its characteristics (e.g. the makeup through petrographic analysis, compressive strength and permeability) and the progression/impact of aggressive species (to the concrete and/or the reinforcement), such as chlorides, sulphates, and carbon dioxide

In the context of ageing infrastructure, the key aim of sampling and testing is to help fill information gaps, as mentioned earlier, as well as to assess the current condition in a more detailed manner and use the results as a base to predict future deterioration. For instance, in order to determine the remaining life of a timber pile in a marine environment, together with the type of interventions potentially suitable to increase its Service Life, materials characteristics, such as the timber species and its load capacity, need to be fed into a structural appraisal model. The risk in that instance of not having this information could lead to decisions being driven by the remaining pile diameter alone, which may be either conservative or optimistic; that is, the margin of error could be such that the confidence level could be low on the likely remaining life, type and timing of any future actions.

At times it may not be possible, practicable or desired to undertake any destructive testing, and thus samples may be collected from the surrounding environment in an attempt to estimate the likely condition of the asset and its remaining life. This can be the case for steel piles where high corrosion rates in the low seawater zone could suggest an aggressive form of Microbiologically Influenced Corrosion (MIC), referred to as Accelerated Low-Water Corrosion or ALWC (Beakell et al., 2005; PIANC, 2005). In that example, comparative water samples may be collected to help inform changes in composition and potential for existence of certain types of bacterial colonies. In another common example, soil and groundwater sampling as foundations (embedded within soil) are typically difficult to assess directly. The general durability process in that instance could look like:

- A survey of any ground information (if available) and sampling to identify the environmental conditions. For instance, the groundwater could be close to seawater in composition (with high chloride and sulphate ions) in contact with high permeability Actual Acid Sulphate Soils (AASS) (BRE Construction Division, 2005), which are very aggressive conditions for some materials.

- Estimate the materials characteristics that are based on the information provided (or experience if no information is available). This could be a concrete pile made with an Ordinary Portland Cement (OPC) mixture and 50 MPa (as designed twenty-eight-day compressive strength) and 50 mm cover to reinforcement driven into the ground in the 1970s.

- Identify the key mode(s) of failure for the materials in the environment. In the aforementioned example, it would be a combination of chloride (potentially the first mode from a kinetic perspective), sulphate and acid-induced deterioration. The deterioration pattern could be the diffusion of chlorides through the concrete toward the reinforcement, followed by a deterioration front where the concrete matrix itself is being degraded.

- Estimate what the Design Life of the piles would have been using different methods, in general application of ISO 13823 (ISO, 2008) and ISO 16586 (ISO, 2001–2011). This involves considering it pertaining to design standards, undertaking deterioration modelling using historical information/evidence or a combination thereof. In the example, the exposure classification would be rated as 'very severe' for concrete piles in soil according to AS 2159 (Australian Standards, 2009a). This standard states that the piles (given the concrete strength and cover) were designed for a one hundred-year life. In 2020, for instance, the remaining life of the pile would be fifty years. However, undertaking chloride diffusion modelling for the OPC mix (as opposed to a marine mix that contains supplementary cementitious materials for increased resistance to chloride ingress and sulphate attack) may suggest a lower remaining life, which allows for two scenarios being considered when identifying future asset management actions.

EMBEDDED HARD-WIRED OR WIRELESS PROBES

Attempts have been made to embed probes into structures to monitor their condition; however, there has not been any systematic adoption. Corrosion 'ladders', for example, utilised to measure ingress of corrosive chloride into concrete, have been made available for insertion into concrete either at the time of construction or post-repairs. These sensors can enable the detection of changes (e.g. the ingress of aggressive species) thus providing advance notices that allow for proactive and more cost-effective management of assets. Considering the risks and costs associated with the management of infrastructure having the ability to monitor ageing of assets as early as possible, this approach has merit and significant potential applications.

In addition to more cost-effective asset management, this approach could positively impact resource usage over the life of the infrastructure, minimise social impacts (e.g. disruptions liked to maintenance activities such as road/tunnel closures) and assist in enhancing the design of future assets.

The development of such technology has challenges. The embedding of a monitoring system within a material at the time of construction raises the question of the life of the system itself (the sensors and their power and communication systems). With long-life assets (such as one hundred years for reinforced concrete tunnels) that stay in a 'good to fair condition' for the vast majority of their life, this could mean having a monitoring system that does not produce very useful data (i.e. confirming that little is happening) for a long time. Given the typical life of electronics, it could also mean that the sensor systems need to be replaced multiple times over the life of the assets. Such replacement could lead to concrete being cut out to extract the probe followed by placement of new concrete around the probe, which would be the same situation for the placement of a sensor within an area of concrete repair. This would mean that the sensor would not be monitoring the condition of the parent/unmodified concrete and thus raise questions on representativeness. The design, deployment and management (including inspection and maintenance) of the monitoring system and its components needs careful considerations, as a system failure could significantly impact the timing and type of remediation interventions to the structure. The embedment of a corrosion ladder into reinforced concrete (that measures the likely corrosion potential and rate at certain depths of cover) requires adequate calibration so that measurement can be relied upon. Incorrect outputs suggesting early onset of corrosion could trigger a decision to repair what is still good undeteriorated concrete around the reinforcement.

COMPILING HISTORICAL CONSTRUCTION/ MAINTENANCE/CONDITION/TEST DATA FROM STRUCTURES

As has been stated, the compilation of data across the asset life cycle that can be turned into information and knowledge/insights is critical to enable optimised and sustainable asset management. Too often, only as-designed drawings are available (and with no documentation in rare cases) for the assessment of existing structures. This leads to not only work having to be undertaken to source information (on and off site) but also presents a risk associated with its reliability. For instance, elements buried in soils will rarely be exposed for inspection due to access limitations or desire not to disturb the ground for environmental reasons. As such, relying on as-designed drawings instead of *as-built* drawings could lead to incorrect assumptions as the basis for the management of the asset. There has been an increasing awareness of the importance of the collection and management of life-cycle data, which feature pre-eminently in leading asset management documents such as 'Asset Management: An Anatomy' (IAM, 2015), the IIMM (IPWEA, 2015), and/or ISO 55001 (ISO, 2014). This would be supported by the use of tools such as Building (read Asset) Information Modelling type technology that allows for full life cycle approach (HM Government, 2012).

Ideally the basis of durability design allows for a baseline 'deterioration curve' to be created, against which future assessments of condition (e.g. inspections, condition surveys, and test results) can be compared. This enables the realisation that the Service Life is likely to well exceed the Design Life with minimal maintenance requirements, or on the other hand provide warnings of the potential need for early interventions due to more advanced ageing. The adequate recording of information as the infrastructure deteriorates is also critical. Site condition data can be presented in so many formats ranging from very long pdf-type reports, MS Excel type spreadsheets or databases, to representation through Geographical Information Systems (GIS). The datasets themselves can vary greatly in terms of attributes (e.g. the number and type of fields recorded), structure (e.g. different asset hierarchy followed, if any), use of free text versus numbers, recording of objective and subjective data, geo-referencing for future location and comparison of assets and defects. Metadata (data on data) can also differ significantly ranging from none (no indication on data quality, reliability, reproducibility of testing) to advanced. These challenges can lead to significant time, efforts and costs being spent on manipulating the data and performing basic/limited analysis as opposed to supporting more advanced analytics that gains valuable insights, thus allowing for more informed decisions being made.

The management of information, particularly as it transitions from the stages of planning to design to construction and (possibly the most important one) construction, is essential. As not one IT system is likely to be handling all the required information (e.g. it could be spread between GIS, drawing sets in CAD, asset registers, work management systems or building information modelling), it is key that the technology environment is set to enable the required level of connectedness and interactions.

RISK PROFILING BASED ON HISTORICAL DATA

Risk-based decision-making

Risk management is at the core of asset management and a key process to determine the scope and timing of any health monitoring regime. Reliability-centred maintenance is considered good practice to enable risk-based decision-making (IAM, 2015; IPWEA NAMS, 2015). This approach enables the tailoring of the scope, timing and frequency of inspections and interventions to provide the desired outputs for the asset manager. The availability of reliable historical data can be a key enabler of the optimisation of inspection and maintenance programs. On the other hand, absence or low quality of reliable historical data can increase the risk of making inadequate consequent decisions and thus the follow-up costs that need to be spent to manage this risk. For instance, this may mean having to undertake a significant audit of all the organisation's assets to collect base information (e.g. location, age, materials of construction, configuration and geometry) before being able to identify any opportunity for more tailored and/or advanced inspection/testing regime.

The accurate and reliable collection of historical information can support trend analysis that feeds into a potentially probabilistic quantification of the likelihood of asset failure. Such an approach, however, needs to be strategically aligned with the owner's objectives and capability, as well as fit for purpose. This is not to say that qualitative assessment of risks based on the criticality of the assets and decision-making process cannot be completely satisfactory and adequate but rather the type of analysis needs to be deemed fit for purpose/appropriate for the organisation. In some instances, a probabilistic approach can enable the costing of the risk (of failure or defect from deterioration) that then can be compared to intervention cost(s) to help assess the benefits of adopting a particular intervention strategy. As an example, the historical trend of condition data may suggest that the probability of asset failure is less than 10 percent over the next ten years. If the consequence of asset failure is estimated to be 'X' dollars, corresponding to the costs of reactive repairs including access, overheads, etc., the risk cost would be less than $0.1X$. Such risk may be below the allowable risk of the asset owner and thus be considered acceptably managed under the current maintenance regime. If the probability of failure had been higher, the increased risk cost may have triggered a review of the current approach and the decision to intervene by remediation in the short to medium term to lower the risk. Such considerations are discussed in the following section.

INTERVENTION STRATEGIES

The competing pressures in many economies across the world for the funding of new and/or the renewal of infrastructure has driven the need to optimise intervention strategies, including Service Life extensions, greater return on investment, lower annualised costs, lower disruption and/or environmental footprint, so as to achieve better outcomes from the existing assets. This has required an increased focus on monitoring the condition of assets, understanding their modes of failure (i.e. failure pathways) and developing tools and processes to better assess and predict their current and future performance respectively (Blin et al., 2008).

In order to maintain the desired serviceability over the asset life, or greater performance through increased Service Life, the wider strategic business objectives need to be connected to tactical decisions and operational considerations. This can be done by supporting infrastructure management processes with asset and 'people' knowledge that includes structural and materials engineering considerations (including of key ageing mechanisms/modes of failure) as previously discussed. This enables the identification of options (from 'do nothing' to complex technical solutions) that can then be assessed by decision-makers in terms of the potential performance, cost and risk impact to the business and its stakeholders.

Decision-making frameworks and supporting tools of varying complexity exist and are used by organisations based on their needs and level of maturity in asset management. For instance, some may adopt an intervention strategy based on a 'run to fail' approach, where repairs will only be undertaken when it is assessed to be necessary based on condition. This reactive approach may be acceptable for low criticality assets but could prove costly and risky when it has the potential to significantly impact infrastructure users. In some strategies, asset condition alone triggers interventions (that can be prescribed as the onset in a manual or electronic database). As this does not consider the consequence of asset failure as previously discussed (and thus infrastructure criticality), this could lead to unnecessary repairs to be carried out and/or a suboptimal use of funds that could otherwise be better spent on more valuable interventions elsewhere. Such consideration is very important when funding is constrained and an organisation has to prioritise works across a wide (in terms of number, types and/or locations) asset base. An example of this would be to launch into the patch repair of a reinforced concrete bridge deck located over water, based on its visual condition without knowing that significant loss of reinforcement in the deteriorated areas can actually be permissible (i.e. there is no urgency to intervene), whereas the guardrails on a different bridge are heavily corroded to the point of not providing the required safety protection. In this instance, the guardrails are more critical to the service delivery and thus should be repaired/replaced first.

Infrastructure criticality can also be used as a key driver of intervention strategies. For instance, the inspection and maintenance program of a critical asset could be more extensive than less important assets to the point that repairs are undertaken prematurely or in preference over any other assets. While this presents the advantage of prioritising works based on the importance of the infrastructure on the service delivery, it can also lead to overservicing critical assets while letting noncritical ones deteriorate to the point that they require significant repairs or replacement (and the opportunity for proactive/preventative treatments is missed). As an example, water reservoirs tend to be critical and reliable assets with long service lives. Prioritising all condition assessment and repair works to these important civil assets over, say, the piping and associated valves coming in/out of the reservoirs, may cause service disruption when these components are found to be deficient as they need to be operated.

Thus, a more advanced approach combines likelihood (e.g. from condition data) and consequence of asset failure to obtain a risk. This, in turn, drives the performance and condition monitoring programs, as well as interventions. A risk-based approached is at the core of Reliability Centre Maintenance and supports the balancing of priorities as the infrastructure ages, the level of funding, and the acceptability of risk appetite of the asset manager/owner. The intervention strategy could, for example, consist of running low-risk assets to failure, while high-risk assets have tailored monitoring regimes and resilience plans allowing them to remain serviceable while being repaired and/or efficiently returned to service after a damaging event.

Risk considerations alone may not allow for the optimised prioritisation of interventions, particularly in a funding-constrained environment that has become relatively common. This is where intervention strategies can be driven by considerations of balancing life-cycle risks and costs. An example of such an approach on a marine jetty could be as follows:

- A jetty has a replacement value of $X and a probability of failure based on its condition of 75 percent within the next five years; that is a risk cost of $0.75X.
- As the asset owner cannot afford to replace the asset, it is considering two options:
  1. The repair and protection of all crossbeams, which would reduce the probability of failure down to 10 percent (a reduction in risk costs of $0.65X) for a cost of $0.3X (a 1:2 ratio between the sum spent and the risk reduction).
  2. The repair and protection of selected beams (including all critical ones) only, leading to a risk reduction of $0.5X (a probability of failure of 25 percent) for a cost of $0.165X (a ratio cost to risk reduction close to 1:3).
- As the organisation may only be able to fund $0.2X of repairs, it could select option (2) and use the balance of funds to apply protective treatment or protection to the elements in the next risk category.

Building on the previous approach, considerations for prioritisation can be extended to balancing performance, risk and cost. In the example, service provided by the infrastructure can be modified to enable an acceptable level of risk for an affordable cost, illustrated by reducing the allowable loadings on a jetty to defer or reduce the scope of repairs. Using the jetty example, the risk of failure could be reduced by modifying the operational areas where vehicle loading would be restricted to a specified corridor. This, in turn, would reduce the load requirements of a number of piles, whose lower utilisation would remove the need for their repairs. The resulting outcome of this performance requirement change would be a reduction in risk and cost to an acceptable level for the organisation (but reduced functionality of the jetty).

This approach can be applied at a component, asset, asset systems and asset portfolio level, which can bring about a high degree of complexity. The output of the prioritisation process can lead to the development of an intervention strategy focused on a number of similar assets (or their components) across geographies. An example would be safety handrails for visitors viewing platforms being repaired or replaced as part of a program of works that brings about consistency in quality of workmanship, accountability on timeliness and economy of scale potentially. Such a complex multilayered prioritisation process would typically be undertaken by mature organisations, which would use asset management electronic tools/ algorithms that support optimised decision-making.

The types of available interventions that can form parts of strategies can vary, typically based on the following considerations:

1. 'Do nothing' and keep monitoring: this is the baseline/default option against which other scenarios can be compared in terms of performance, risk and cost differential. This would be often used for low-risk assets/elements and within a short to medium time period.
2. Reactive or planned intervention:
   - Strengthening, upgrade: if the structure is found to be deficient (e.g. high risk of not being able to carry the required loads), works can be designed to increase its functionality (provide greater levels of service). Such an option can include carbon fibre strengthening of reinforced concrete elements, installation of bracing elements, driving of new piles, installation of reinforced concrete jackets on timber piers, etc.
   - Rehabilitation, repair: where the asset is still structurally adequate but at risk of not being so in the near future, and/or where its condition poses safety/environmental concerns repairs can be triggered. This can take many forms depending on the Design Life requirements: repairs can be relatively short term (for a few years to allow for the planning of major/renewal works), medium term (e.g. fixing current defects with repairs that are expected to last the ten to fifteen years it is estimated it would take for areas not currently damaged to require intervention) or longer term (e.g. reinstating the asset to its near original condition to achieve a thirty-plus-year Service Life extension).
3. Preventative works:
   - Early proactive intervention: there are instances where the opportunity exists to prevent/slow down ageing and achieve lower

life-cycle costs and risks than if the assets were left to deteriorate at their current rate. An example would be the application of
protective treatment on steel or reinforced concrete elements that
provides the first barrier/protection layer and thus extends the life
of the assets.

• Future proofing: when predicted future infrastructure demands
require for an asset to provide greater capacity/level of service
than it was designed for, there is an opportunity to plan for works
that strengthen and, if necessary, increase the durability of the
structure for a required life.

The aforementioned considerations are illustrated in the corrosion management for reinforced concrete deterioration that was presented by Blin and
Christodoulou (2014). Chloride-induced and carbonation-induced corrosion are key drivers of ageing for reinforced concrete maritime/coastal and
urban (including road/rail tunnel) assets respectively. As a result, corrosion
management has become a growing subset of asset management over the
past several decades. This has allowed asset managers and their partners
(including consultants and contractors) to identify and tailor intervention
strategies for reinforced concrete structures based on the following options:

• Monitoring including: visual, testing (nondestructive and sampling), prediction analysis and forecasting of deterioration
• Repair such as localised patch repairs or full replacement of parts/
elements
• Surface applications, for example, corrosion inhibitors, coatings,
membranes and hydrophobic impregnation
• Electrochemical treatment consisting of impressed current
cathodic protection (ICCP), Galvanic cathodic protection, chloride extraction, re-alkalisation, reverse osmosis and hybrid corrosion protection
• Design avenues that include replacement, enclosure (to prevent
deteriorated pieces to fall onto public areas for instance), structural modifications

Typically, monitoring and concrete patch repairs form part of short- to
medium-term strategies. These are commonly used, possibly because condition assessment manuals focus on visual inspection with a link to base
interventions, with less emphasis on sampling, testing or broader considerations of performance, risk and cost as discussed previously.

Electrochemical treatments, and particularly cathodic protection (CP),
are typically medium- to long-term strategies and have become very popular
when the Service Life of the assets is required to be extended for twenty years
or more. Design solutions tend to be used as part of long-term strategies and
could include making parts of an asset redundant, modifying its functionality

or strengthening it. Corrosion management strategies can be selected and tailored to the required additional life required for an asset, taking into account the ability to inspect and maintain the structure over time. For instance, the requirement for a fifty-year-old building facade in close proximity to the seashore to remain serviceable can be achieved through different rehabilitation options such as:

1. **Do nothing** until the level of service required can no longer be provided. This would have the lowest cost but possibly the highest risk depending on factors such as the asset's condition, ability to regularly and accurately monitor the asset while in-service, importance of continuous operations on revenue, etc.
2. **Patch repair** only deteriorated areas as they are identified during the scheduled inspection program. This solution would attract some cost while decreasing risk over a do-nothing option that would accumulate over time. The frequency of repairs would be linked to the deterioration rate of both the nonrepaired and repaired areas.
3. **Implement ICCP** across the areas of the structure more prone to significant deterioration affecting its structural integrity. This would typically require a greater investment of funds over a conventional patch repair with the trade-off being a greater Service Life and a lower risk of failure/service interruption.

Each of these options can be considered in their own right, and the selection of one over the other will depend on the organisational framework and its objectives for the sustainable management of its service delivery and thus its assets. Further guidance can be provided in documents such as ISO 16311-1 Maintenance and repair of concrete structures (ISO, 2014). In instances, complex decision trees (often supported by electronic tools) can assist in identifying the potential technical solutions and their risks/benefits to help determine the optimal way and time to renew or to continue to maintain.

CASE STUDY: VIETNAM BRIDGE MAINTENANCE PROJECT

Background

Author F. Collins was involved with setting up a Bridge Testing and Assessment Unit (BTAU) within the Vietnamese Ministry of Transport (Anderson et al., 1998). The single carriageway of the country's main highway, National Route 1 (NR1), linking Hanoi to Ho Chi Minh City is intersected by many rivers and streams and has to traverse several mountain passes. A significant section of the route runs within a kilometre of the coast and is subject to the corrosive effects of the saline environment, as well as the annual typhoon season and the effects of flooding and scouring. Bridges tend to be a weak point of a transport network. A number of bridges were dilapidated due to their age, design, construction and lack of regular maintenance. The upgrading and rehabilitation of NR1 was seen as being a very important element of the transport system. As part of an Overseas Development Administration (UK) aid project, a BTAU was established within the Scientific and Technological Research Institute of the Ministry of Transport (STICT/MoTC), to carry out bridge assessments for part of the subsequent phase of Vietnam's road rehabilitation programme and to develop bridge repair and maintenance techniques and a bridge management system. The BTAU tasks included:

- Training engineers and technicians to undertake bridge assessment and bridge testing
- Introducing techniques of bridge maintenance and repair and adapting such techniques to the conditions in Vietnam
- Developing a bridge inventory and database system suitable for use in a bridge management system (BMS)
- Evaluating the condition of 117 bridges on the section of NR1 between Vinh and Dong Ha to recommend a repair and replacement strategy for the next phase of the government's highway rehabilitation programme

Bridge inspections

Routine *General* Inspections (GI) were visual inspections conducted annually from ground and deck level and from any fixed walkways or travelling gantries built into the structure. Binoculars were required to inspect inaccessible parts.

Routine *Principal* Inspections (PI) had to be undertaken every five to ten years and required close examination within touching distance of all parts of the bridge, complemented by nondestructive testing at selected locations. Suitable access was needed to allow this to be done.

Special Inspections (SI) were required when a problem was found during a routine General or routine Principal Inspection, necessitating targeted investigation. Special Inspections would also be required after flooding, after severe accidental damage and after the passage of an exceptional load.

Manuals covering each of the aforementioned inspection types were developed as a basis for future BTAU bridge inspections and the training of engineers. The manuals covered not only the background theory and technical details required in the inspections but also the systematic means of recording the results of the inspections and the analysis of these results. Examples of bridges inspected are shown in Figure 7.2.

Figure 7.2 (a) Barrel vault arched bridge; (b) Bridge with concrete deck on structural steelwork showing drying shrinkage cracking; (c) Steel superstructure and concrete deck; (d) Reinforced and prestressed concrete bridge. (Photographs by F. Collins.)

Bridge assessment

Evaluation of load-carrying capacity of bridges formed an important activity in the overall process of bridge management. Bridge assessment was undertaken to determine, in terms of vehicle loading, the load that a given structure would carry with a reasonable probability that it would not suffer serious damage so as to endanger any persons or property on or near the structure. The method of bridge assessment adopted was to use structural calculations, supplemented by inspection and in situ testing to establish material properties, condition and structural integrity. A bridge load test was not an alternative to this form of assessment and should have only been considered after the calculations had been completed when the assessment was considered unsatisfactory.

Bridge management system

The bridge management system (BMS) was developed specifically for use in Vietnam by the Regional Road Management Units (RRMUs) as a means of assisting them in the management of their bridge stock. The *BridgeMan* bridge management system developed by Oxfordshire County Council in the United Kingdom was selected from a number of available systems, as it was considered to be the most suitable for use in Vietnam. *BridgeMan* took advantage of Windows graphical interface and included the use of a Vietnamese keyboard, thus allowing for a full Vietnamese language version of the system. The work in developing the *BridgeMan* (Vietnam) BMS was initiated by Oxfordshire County Council in conjunction with the UK's Transport Research Laboratory and was completed in Vietnam by project team members from the BTAU. The key aspects of the BMS were to be a repository of records and a decision tool including bridge inventory information, design and construction data, updated with the inspections GI, PI, and SI, together with the structural bridge assessments, records of maintenance and rehabilitation and decision support. The data generated by the inspections of the 117 bridges between Vinh and Dong Ha was used to populate the BMS. A Vietnamese Language User Manual was written by members of the BTAU to facilitate the use of *BridgeMan* (Vietnam) both along NR1 and elsewhere in Vietnam.

Training

The overall objective of introducing techniques of bridge maintenance and repair required the project to undertake a series of training courses and workshops held both in Vietnam and the UK. These training courses were coupled with both practical training on site and by the undertaking of work, which could contribute to Vietnam's ongoing programme of bridge rehabilitation. Two of the major objectives of the project were the transfer of technology and the institutional strengthening. Thus, in addition to the more formalised workshops and training courses, ongoing training took place throughout the project. This happened whenever either the local consultants or counterparts worked with other members of the project team.

Inspection and assessment of bridges from Vinh to Dong Ha

The project brief required application of the techniques of bridge inspection and assessment developed with the Vietnamese counterparts, to the

117 bridges between Vinh and Dong Ha, along NR1. Thus, this component of the project had two aims:

- The application of training techniques to actual inspection and assessment
- The development of a strategy for bridge repair and replacement

The inspections of 117 bridges were undertaken in a three-week period. The inspections included both a bridge inventory and a visual inspection of all visible parts of the structures. The inspections were carried out in accordance with the Project's GI and PI manuals with the completion of the standard inventory and inspection forms and the production of an inspection report (including sketches and photographs) being produced for each of the inspected bridges.

From the findings of the GIs, three bridges were selected for PIs and SIs on the basis of:

- Apparent condition of structure from GIs
- Ease of access for relatively large numbers of trainers and trainees
- Types of problems likely to be encountered
- Structural type of the bridge

The one overriding condition in this selection was to give as wide a range of experience as possible to the Vietnamese engineers undertaking PIs and SIs and the subsequent assessments. Following inspection, 'bridge assessment' included both structural and condition appraisal and determination of the load-carrying capacity of a structure and the maintenance on its ability to function satisfactorily in the long term.

Repair and strengthening

As a natural follow-up to bridge inspection and bridge assessment, work was then undertaken with Vietnamese counterparts on the development of suitable methods to repair and strengthen the bridges. It was only possible to specify suitable repair and strengthening techniques to a bridge following completion of a PI and a full assessment. Through the inspections of Vietnamese bridges, many common defects were detected, and the current methods of repair and strengthening were reviewed. Through both formal and informal training sessions, current and new procedures for repair were identified and these procedures were documented in the project's bridge repair manual. A framework procedure for prioritising rehabilitation and maintenance on the bridges on the Vinh to Dong Ha route was developed. Repair priority was directly related to the consequence of failure of a particular bridge element. That was, if the consequence of failure was high (i.e. it would affect public safety or continued use of the bridge), then the repair priority also had to be high. After repairs had been prioritised on any bridge stock, the exact type of repair necessary for each individual bridge had to be determined. This was only possible following a full bridge assessment and consideration of other nontechnical factors. These factors included:

- Required Service Life of bridge
- Available budget (for the capital, maintenance or life-cycle cost)
- Available manpower resource
- Allowable disruption during bridge repairs
- Aesthetics of repaired bridge

Durability for new construction

As previously mentioned, the ability of a structure to retain its serviceability during its Service Life and its durability is considered to be dictated by the following factors:

1. The specification of materials used in construction
2. The design and detailing of the structure
3. The type and quality of construction practice

Each of these factors was addressed during the term of the project through both formal and informal training sessions to ensure knowledge transfer and a feedback loop for the creation of new assets. By ensuring that bridges were then designed with future durability issues in mind, lower whole-life costs could be achieved, as there would be less need for subsequent maintenance and repair.

CONCLUSION

This chapter provides an overview of the rationale and considerations for the monitoring of infrastructure health and associated intervention strategies. It emphasised the importance of understanding the organisational need for performance and condition monitoring, as well as how it should be tailored to the mode(s) of asset failure and associated risks. It also outlined the importance for considerations of balancing asset performance, risk and cost to drive the development of intervention strategies so they provide best value to organisations and their stakeholders.

REFERENCES

AECOM. (2012). Enhanced asset management using intelligent monitoring. In: E. Sullivan and P. Davis (Eds.), *Report for Intelligent Water Networks Program as Part of Network Efficiency Project.* https://vicwater.org.au/wp-content/uploads/2013/01/IWN_Network-Efficiency_AECOM_Enhanced-Asset-Management-Using-Intelligent-Monitoring_Report_Final.pdf.

Anderson, G., Collins, F., and Grace, W. (1998). Development of bridge evaluation and maintenance management methodologies for the road network of Vietnam. In: D. W. S. Ho, I. Godson, and F. Collins (Eds.), *Proceedings of 2nd RILEM International Conference on Rehabilitation of Structures.* Melbourne, Australia, pp. 577–591.

Australian Standards. (2005). *AS/ISO 13822, Basis for Design of Structures—Assessment of Existing Structures.* SAI Global, 36p.

Australian Standards. (2009a). *AS 2159 Piling—Design and Installation.* SAI Global, 85p.

Australian Standards. (2009b). *AS/NZS ISO 31000 Risk Management—Principles and Guidelines.* SAI Global, 16p.

Australian Standards. (2009c). *AS/NZS ISO 31000 Risk Management—Principles and Guidelines.* SAI Global, 37p.

Beakell, J. E., Foster, K., and Siegwart, M. (2005). *CIRIA C634 Management of Accelerated Low Water Corrosion in Steel Maritime Structures.* CIRIA, 29p.

Bertolini, L., Elsener B., Pedeferri, P., and Polder, R. B. (2013). *Corrosion of Steel in Concrete: Prevention, Diagnosis, Repair,* 2nd ed. Weinheim, Germany: Wiley, 392p.

Blin, F., Law, D., Dacre, M., Hoog, C., Gray, B., and Newcombe, R. (2008). Extension of design life of existing maritime infrastructure—A durability perspective. In: W. Mandeno (Ed.), *Proceedings of the Corrosion & Prevention Conference,* 16–19 November 2008. Wellington, New Zealand: Australasian Corrosion Association, pp. 1–13.

Blin, F. and Christodoulou, C. (2014). Increasing the durability of existing infrastructure assets, *International Congress on Durability of Concrete.* New Delhi, India, 8p.

BRE Construction Division. (2005). *BRE Special Digest 1—Concrete in Aggressive Ground,* 3rd ed. Watford, UK: BRE Press, 62p.

Buenfeld, N. R., Davies, R. D., Karimi, A., and Gilbertson, A. L. (2008). Intelligent monitoring of concrete structures (C661), *Construction Industry Research and Information Association (CIRIA),* 126p.

CIRIA. (2006). In: C. Melbourne, L. D. McKibbins, N. Sawar, and C. Sicilia Gaillard (Eds.), *C656 Masonry Arch Bridges: Condition Appraisal and Remedial Treatment.* Construction Industry Research and Information Association (CIRIA), 342p.

CIRIA. (2010). In: L. McKibbins, R. Elmer, and K. Roberts (Eds.), *PUB C671 Tunnels: Inspection, Assessment and Maintenance.* Construction Industry Research and Information Association (CIRIA), 247p.

Concrete Institute of Australia. (2015). *Z7/07 Performance Tests to Assess Concrete Durability.* SAI Global, 97p.

HM Government (2012). Industrial strategy: Government and industry in partnership, Building Information Modelling. https://www.gov.uk/government/uploads/system/uploads/attachment_data/file/34710/12-1327-building-information-modelling.pdf.

Institute of Asset Management. (2015). Asset management: An anatomy, International Infrastructure Management Manual (IIMM), Institute of Public Works Engineering Australasia (IPWEA). https://theiam.org/knowledge/Knowledge-Base/the-anatomy/.

IPWEA NAMS. (2015). CIRIA was in the word document provided: Construction Industry Research and Information Association or CIRIA 2008, Intelligent monitoring of concrete structures (C661).

IPWEA NAMS. (2015). Condition assessment and asset performance guidelines, practice note 3: Buildings. Institute of Public Works Engineering Australasia (IPWEA), 26p.

International Standards Organisation. (2001–2011). *ISO 15686, Parts 1–9 Buildings and Constructed Assets-Service Life Planning.* Geneva, Switzerland: SAI Global.

International Standards Organisation. (2008). *ISO 13823 General Principles on the Design of Structures for Durability.* Geneva, Switzerland: SAI Global, 39p.

International Standards Organisation. (2014). *ISO 16311-1 Maintenance and Repair of Concrete Structures.* Geneva, Switzerland: SAI Global, 19p.

PIANC. (1990). *Report 17, Inspection, Maintenance and Repair of Maritime Structures Exposed to Material Degradation Caused by a Salt Water Environment, II,* Working Group 17, PIANC, 26p.

PIANC. (2005). Accelerated low water corrosion. *Permanent International Association of Navigation Congresses.* Permanent Technical Committee, Working Group 44, PIANC, 32p.

Ports Australia. (2014). Wharf structures condition assessment manual. http://www.portsaustralia.com.au/assets/Publications/Wharf-Structures-Condition-Assessment-Manual-About.pdf.

VicRoads. (2010). Technical note 61 non-destructive testing (NDT) of concrete in structures. https://www.vicroads.vic.gov.au/business-and-industry/technical-publications/pavements-geotechnical-and-materials.

VicRoads. (2014). Road structures inspection manual. https://www.vicroads.vic.gov.au/business-and-industry/technical-publications/bridges-and-structures.

WERF. (2007). Condition assessment strategies and protocols for water and wastewater assets, Water Environment Research Foundation. https://www.werf.org/a/ka/Search/ResearchProfile.aspx?ReportId=03-CTS-20CO.

Chapter 8

The future

INTRODUCTION

Future action on ageing infrastructure will be driven by social, economic and environmental imperatives. The economic and environmental drivers are self-evident; however, the social drivers demand more flexible, reusable and community-focused infrastructure. The way that we undertake action on ageing infrastructure will be governed by new technologies evolving for more durable construction, methods of surveillance of the condition of built infrastructure, and information systems that provide a more proactive approach toward the construction and in-service condition and management of infrastructure. This chapter will explore these themes.

ECONOMIC NEEDS OF BUILT INFRASTRUCTURE

The American Society of Engineers (ASCE) recently undertook a comprehensive evaluation of the condition of built infrastructure (ASCE, 2017). The report concluded that deteriorating infrastructure is impeding the ability of the United States to compete in the thriving global economy, and improvements are necessary to ensure the United States is built for the future. Overall, the condition of America's infrastructure is rated D+ or 'The infrastructure is in poor to fair condition and mostly below standard, with many elements approaching the end of their Service Life'. Failing to close this infrastructure investment gap ($2.0 trillion) to bring infrastructure from the current rating of D+ to B brings serious future economic consequences: $3.9 trillion in losses to the US GDP by 2025; $7 trillion in lost business sales by 2025 and 2.5 million lost American jobs in 2025.

In Europe, the quality of infrastructure has deteriorated due to the ageing of networks and insufficient maintenance spending (European Commission, 2014). Underspending in maintenance within the road, rail and energy sectors has led to deterioration of the quality of the network, hence, lowering the efficiency of the whole network. Maintenance of ageing infrastructure includes different types of quality enhancements, such as local repair, winter maintenance, renewal, and addition of new functionalities (bridge, tunnel, etc.), as well as prolongation of the lifetime of existing infrastructures. Overinvesting in new infrastructure is associated with underspending on maintenance – whether this overspending on new infrastructure translates to more resilient and durable infrastructure is not clear.

While traditionally initial capital costs have been the predominant consideration, there is a realisation that future infrastructure spending is needed for maintenance/rehabilitation, and resource allocations are needed to control ageing of infrastructure.

ENVIRONMENTAL IMPACTS

Construction of new infrastructure and the activities associated with the maintenance and rehabilitation are energy-intensive activities. Globally, concrete is the predominant construction material, exceeding 25 gigatonnes/year (Gursel et al., 2014), and the activities to produce this construction material are energy intensive. For example, manufacturing Portland cement is very energy intensive, while the calcination of limestone during manufacturing leads to high CO_2 emissions. Future construction methods will seek to use less energy-intensive raw materials while seeking lower-energy methods of construction, maintenance and rehabilitation. More durable built infrastructure will reduce ongoing maintenance and rehabilitation activities during the life cycle, thereby reducing CO_2 emissions of ageing infrastructure.

Future global warming, leading to higher ambient temperatures, will increase the kinetics and rate of deterioration of infrastructure. Increased solar and terrestrial radiation will increase the weathering of exposed polymers and plastics. Increasing greenhouse gas concentrations will lead to higher acidity of the air in contact with exposed infrastructure, thereby leading to higher corrosivity.

Wang et al. (2010) analysed the climate change impacts on the future deterioration of Australian reinforced concrete infrastructure. Most designs of new infrastructure do not consider the effect of a changing environment. Wang et al. (2010) undertook a probabilistic simulation approach based on chloride- and carbonation-induced corrosion models, which were applied to simulate the effect of climate change. The study concluded that increasing carbon dioxide levels will increase the carbonation and thus enhance the likelihood of carbonation-induced corrosion. Meanwhile, elevated local temperatures due to global warming will accelerate the ingress of corrosives into concrete, as well as increasing the rate of deterioration of carbonation-induced and chloride-induced reinforcement corrosion. The greater incidence of extreme events (such as flash floods, bushfires, and strong winds) will impact the infrastructure more intensely and frequently, leading to greater ageing.

NEW CONSTRUCTION MATERIALS AND METHODS

Already, the industry is making inroads toward lower energy materials. Partial replacement of Portland cement with waste by-products (e.g. ground granulated blast furnace slag, fly ash, silica fume, and metakaolin), as well as alkali-activated binders (referred to as 'geopolymers' and based totally on waste by-products and therefore zero Portland cement) will reduce the carbon footprint, as well as provide cost economies and better durability of concrete.

The advent of future construction materials (new composites, memory retaining alloys, smart materials, and nanomaterials) has the potential toward being lower energy materials to produce and install, while also providing improved durability. Smart materials provide opportunities for self-sensing built infrastructure, including, for example, the microclimate, corrosivity, structural properties and loadings. Bacteria can be utilised to precipitate solids, whereby calcite is deposited within porous and/or cracked concrete, thereby healing and producing a more durable concrete. Carbon nanotube-concrete composites (Chen et al., 2011) have the potential to provide better tensile strength while reducing the need for a portion of the steel reinforcement, resulting in slimmer, more durable, and less heavy concrete that reduces quantities for construction and providing lower carbon emissions.

Three-dimensional printing of future components of infrastructure has the potential to provide more economical, less energy intensive, and more durable construction. There is potential to develop 3D printing of inspections and repairs on aged infrastructure.

The impact of the built infrastructure has been considered. For example, installing green vegetated roofs and walls on buildings reduces the urban island heating effect, while reducing the needs for internal heating and cooling – which add to carbon emissions – as well as being architecturally more pleasing. Whether this translates toward reducing the ageing of infrastructure is not clear.

CONDITION INTERROGATION AND TESTING

Condition surveys of infrastructure involve visual inspection and nondestructive testing. During operation of infrastructure, condition surveys are infrequent and, depending on the built infrastructure, can cause difficulties of access, Occupational Health and Safety (OH&S) requirements and typically only involve a portion. Smart, noncontact methods ensure regular and ongoing feedback on the durability of built components. Embedded and surface sensors have the capability of obtaining climatic, physical and durability data at regular time intervals, and have the potential to provide both specific and general information. For example, embedded and attached sensors can inform by WiFi the real-time presence, concentration, rate and location of corrosives ingress, which will provide a proactive means of asset management.

Drones and robotic crawling devices can access difficult locations while providing high resolution visual images, as well as noncontact nondestructive testing.

These efficiencies will improve decision-making for maintenance and rehabilitation leading to better infrastructure durability, longer Service Life and improved confidence.

BUILDING INFORMATION MODELLING

Building information modelling (BIM) is a method that creates and manages all of the design, construction and maintenance information over the project life cycle. The physical and functional data and information on a project becomes s a digital item on every component of the built infrastructure. BIM is already being implemented on projects and will be more widely utilised in the future. In terms of management of ageing infrastructure, rather than disparate and disconnected registers of information, the construction, condition inspection and testing, and maintenance information belonging to a particular component can be retrieved from a single source. Durability Plan components (discussed in Chapter 6) will readily translate within the construction and operations aspects of the BIM.

ASSET MANAGEMENT

Managing ageing infrastructure is becoming more focused on data management and quality, as discussed in Chapter 7. Industry is promoting a forward-looking approach, based on risk assessments and smart information management that strives for better prioritisation and decisions. The advent of wireless devices installed at vital locations on the asset, when compared with the current methods of manual inspection, will provide continuous monitoring of the asset condition and allows more proactive planning of future maintenance. Autonomous vehicles, for example, drones or robotic crawling devices utilising artificial intelligence, will provide more comprehensive condition data. How the vast amount of data is managed over the life cycle is critical toward sound management of the ageing asset.

CONCLUSION

The improved reliability of ageing infrastructure will be better managed by a range of options discussed in this book and potential future methodologies. The community, government and industry will benefit from a proactive approach to ageing infrastructure. This will translate to lower disruption of services, such as traffic closures, outages of water and electricity or port disruption. Over the life cycle, ageing infrastructure should provide improved economies due to more efficient management, durable construction and more proactive approaches to maintenance and rehabilitation.

REFERENCES

ASCE. (2017). 2017 Infrastructure report card, American Society of Civil Engineers. https://www.infrastructurereportcard.org.

Chen, S. L., Collins, F. G., Macleod, A., Pan, Z., Duan, W. H., and Wang, C. M. (2011). Carbon nanotube–cement composites: A retrospect, *The IES Journal Part A*, 4(4): 254–265.

European Commission. (2014). Infrastructure in the EU: Developments and Impact on Growth, European Economy, Occasional Papers 203. http://ec.europa.eu/economy_finance/publications/.

Gursel, P. A., Masanet, E., Horvath, A., and Stadel, A. (2014). Life-cycle inventory analysis of concrete production: A critical review, *Cement and Concrete Composites*, 51: 38–48.

Wang, X., Nguyen, M., Stewart, M. G., Syme, M., and Leitch, A. (2010). *Analysis of Climate Change Impacts on the Deterioration of Concrete Infrastructure—Part 3: Case Studies of Concrete Deterioration and Adaptation*. Canberra, Australia: CSIRO.

Index

Note: Page numbers in italic and bold refer to figures and tables respectively.

Milton Keynes UK
Ingram Content Group UK Ltd.
UKHW040051071024
449327UK00019B/487